Chemistry

Students' Book II

Topics 13 to 19

14 082652 1
Nuffield Advanced Science

Project team

E. H. Coulson, formerly of County High School, Braintree (organizer)
A. W. B. Aylmer-Kelly, formerly of Royal Grammar School, Worcester
Dr E. Glynn, formerly of Croydon Technical College
H. R. Jones, formerly of Carlett Park College of Further Education
A. J. Malpas, formerly of Highgate School
Dr A. L. Mansell, formerly of Hatfield College of Technology
J. C. Mathews, King Edward VII School, Lytham
Dr G. Van Praagh, formerly of Christ's Hospital
J. G. Raitt, formerly of Department of Education, University of Cambridge
B. J. Stokes, King's College School, Wimbledon
R. Tremlett, College of St Mark and St John
M. D. W. Vokins, Clifton College

Chemistry

Students' Book II

Topics 13 to 19

Nuffield Advanced Science
Published for the Nuffield Foundation by Penguin Books

Penguin Books Ltd, Harmondsworth, Middlesex, England
Penguin Books Inc., 7110 Ambassador Road, Baltimore, Md 21207, U.S.A.
Penguin Books Ltd, Ringwood, Victoria, Australia

Reprinted 1971, 1974

Filmset in 'Monophoto' Times New Roman by
Keyspools Ltd, Golborne, Lancs,
and made and printed in Great Britain by
Butler & Tanner Ltd, Frome and London

Design and art direction by Ivan and Robin Dodd
Illustrations designed and produced by Penguin Education

Contents

Foreword

Sixth form courses in Britain have received more than their fair share of blessing and cursing in the last twenty years: blessing, because their demands, their compass, and their teachers are often of a standard which in other countries would be found in the first year of a longer university course than ours: cursing, because this same fact sets a heavy cloud of university expectation on their horizon (with awkward results for those who finish their education at the age of 18) and limits severely the number of subjects that can be studied in the sixth form.

So advanced work, suitable for students between the ages of 16 and 18, is at the centre of discussions on the curriculum. It need not, of course, be in a 'sixth form' at all, but in an educational institution other than a school. In any case, the emphasis on the requirements of those who will not go to a university or other institute of higher education is increasing, and will probably continue to do so; and the need is for courses which are satisfying and intellectually exciting in themselves – not for courses which are simply passports to further study.

Advanced science courses are therefore both an interesting and a difficult venture. Yet fresh work on advanced science teaching was obviously needed if new approaches to the subject (with all the implications that these have for pupils' interest in learning science and adults' interest in teaching it) were not to fail in their effect. The Trustees of the Nuffield Foundation therefore agreed to support teams, on the same model as had been followed in their other science projects, to produce advanced courses in Physical Science, in Physics, in Chemistry, and in Biological Science. It was realized that the task would be an immense one, partly because of the universities' special interest in the approach and content of these courses, partly because the growing size of sixth forms underlined the point that advanced work was not *solely* a preparation for a degree course, and partly because the blending of Physics and Chemistry in a single advanced Physical Science course was bound to produce problems. Yet, in spite of these pressures, the emphasis here, as in the other Nuffield Science courses, is on learning rather than on being taught, on understanding rather than amassing information, on finding out rather than on being told: this emphasis is central to all worthwhile attempts at curriculum renewal.

If these advanced courses meet with the success and appreciation which I believe they deserve, then the credit will belong to a large number of people, in the teams and the consultative committees, in schools and universities, in authorities and councils, and associations and boards: once again it has been the Foundation's privilege to provide a point at which the imaginative and helpful efforts of many could come together.

Brian Young
Director of the Nuffield Foundation

Contributors

Many people have contributed to this book. Final decisions on the content and treatment of the teaching scheme with which it is concerned were made by members of the Headquarters Team, who were also responsible for assembling and writing the material for the several draft versions that were used in school trials of the course. During this exercise much valuable help and advice was generously given by teachers in schools, and in universities and other institutions of higher education. In particular, the comments and suggestions of teachers taking part in the school trials have made a vital contribution to the final form of the published material. Acknowledgment for this assistance is given at the end of this book. The editing of the final manuscript was carried out by B. J. Stokes.

At a time when the Headquarters Team was small in number and under heavy pressure, the following teachers undertook between them the task of developing experimental investigations and planning the theoretical treatment for the first drafts of Topics 12, 14, 15, 16, 17, 18, and 19: W. H. Francis and G. R. Grace (Apsley Grammar School, Hemel Hempstead), K. C. Horncastle (Exeter School), D. R. P. Jolly (Berkhamsted School), Dr R. Kempa (College of St Mark and St John), J. A. Kent (Southfield School, Oxford), P. Meredith (Exeter School), A. G. Moll (Plymouth College), A. B. Newall (The Grammar School for Boys, Cambridge), M. Pailthorpe (Harrow School), Miss K. Rennert (Cheney Girls' School, Oxford), T. A. G. Silk (Blundell's School), Dr R. C. Whitfield (Department of Education, University of Cambridge). It is fitting that their contribution should be acknowledged here. Thanks are also due to D. R. Browning, for advice and assistance with regard to safety measures in laboratory work.

E. H. Coulson

Topic 13

Carbon chemistry, part 2

In this topic we shall continue with our study of carbon compounds and we shall be particularly concerned with unsaturated compounds, that is, alkenes, carbonyl compounds, and arenes. The chemistry of another functional group, the amine group $—NH_2$, will also be studied, both on its own and in compounds containing two functional groups.

Finally, as the culmination of your study of carbon compounds, a week's work will be devoted to attempting a multistage synthesis in the laboratory.

13.1 **Unsaturated compounds: the alkenes**

Fractional distillation in a refinery separates the crude oil into a set of mixtures which have different boiling ranges (the fractions). The distillation is followed by chemical processing which has to be controlled so as to vary the relative amounts of the fractions. It requires skilful management to operate a refinery so that, for example, petrol (gasoline) and paraffin (kerosine) are produced in the proportions in which both can be sold.

In the USA the most important refinery product is petrol, but in Europe two fractions are of great importance, petrol (boiling range 30–120 °C) and naphtha (boiling range 120–200 °C). Naphtha consists of straight and branched chain alkanes together with some arenes.

Naphtha is the fundamental raw material or feedstock of the petrochemical industry in Europe, and in 1968 over 4 million tonnes were used in the UK alone, a quantity that was increasing by 20 per cent every year! To convert it to useful compounds naphtha vapour is mixed with half its weight of steam (as an inert diluent) and heated to 900 °C for less than a second. The mixture produced is rich in C_1 to C_4 alkenes, and contains some arenes (which are not affected by the steam cracking process). The relative proportions of the products can be varied by changes in the cracking conditions, but a typical composition would be hydrogen 10 per cent, methane 10 per cent, ethylene 25 per cent, propylene 15 per cent, butenes 10 per cent, and petrol and fuel oil 30 per cent, together with some arenes in the amount found in the original naphtha.

The product of major economic importance from naphtha cracking is ethylene. Some of the products derived from ethylene are indicated in figure 13.1a.

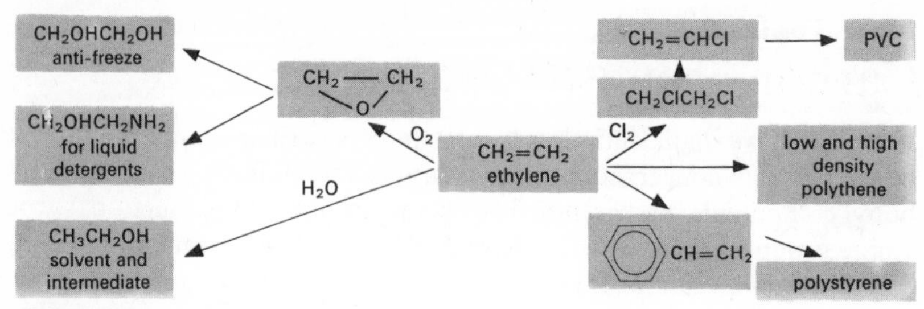

Figure 13.1a
Industrial uses of ethylene.

1	BP refinery	11	ethylene
2	joint power station	12	cumene butadiene
3	polythene	13	ethanol
4	phenol	14	propan-2-ol
5	cumene	15	but-2-ene butadiene
6	ethylene	16	detergent
7	ethylene	17	styrene
8	butadiene	18	gasoline
9	methanol	19	ethylene
10	acrylonitrile	20	ethanol

Figure 13.1b
The Grangemouth works of BP Chemicals (UK) Ltd.
Airviews Ltd, Manchester

The closely related manufacturing processes based on naphtha cracking are usually conducted on one site, as can be seen from figure 13.1b, which is an aerial view of the Grangemouth works of BP Chemicals (UK) Ltd. The ethylene plant in the foreground produces a quarter of a million tonnes each year. The white line across the figure indicates the site boundary. At the top lefthand corner can be seen the BP refinery, and between the two plants the joint power station which supplies both sites.

Propylene (systematic name, propene) is also of considerable economic importance, being used for products such as polypropylene, epoxy resins, and polyurethane foams. You should do some background reading on petrochemicals and attempt to produce your own flow chart for propylene. (See, for example, *Petroleum Chemicals* produced by BP Chemicals (UK) Ltd.)

Summary of the reactions of alkenes

1 *Reduction with hydrogen* – Hydrogen gas will add to alkenes forming alkanes at ordinary temperature and pressure if a catalyst of finely divided platinum is used. Heat and hydrogen under pressure may be needed if a nickel catalyst is used but this is overall the cheaper method, and this catalyst is used industrially. Unsaturated naturally occurring oils can be converted to saturated solids with low melting points (which are useful, for example, in margarine) by this process.

$$CH_3(CH_2)_7CH{=}CH(CH_2)_7CO_2H + H_2 \rightarrow CH_3(CH_2)_{16}CO_2H$$

2 *Hydration to alcohols* – The C_2 to C_4 alkenes are available in quantity in the petrochemical industry and can be readily converted to secondary alcohols by dissolving in moderately concentrated sulphuric acid followed by dilution.

$$CH_3CH{=}CH_2 \xrightarrow{H_2SO_4} CH_3{-}\underset{\displaystyle OSO_3H}{\underset{|}{CH}}{-}CH_3 \xrightarrow{H_2O} CH_3{-}\underset{\displaystyle OH}{\underset{|}{CH}}{-}CH_3$$

Note that in this reaction the secondary alcohol, propan-2-ol, is the product. A reaction which produces primary alcohols is described below.

Direct hydration is also possible; ethanol can be made by the direct hydration of ethylene.

$$CH_2{=}CH_2 + H_2O \rightarrow CH_3CH_2OH$$

A high temperature and pressure, and a catalyst of phosphoric acid, are required.

3 *Addition of halogen hydrides* – If alkenes are treated with concentrated hydrobromic or hydriodic acids, a secondary halogenoalkane will be obtained:

$$RCH{=}CH_2 + HBr \rightarrow \underset{\displaystyle Br}{\underset{|}{RCH}}CH_3$$

Hydrogen chloride requires more drastic conditions for addition to occur.

4 *The 'oxo' reaction* – When alkenes are heated under pressure with carbon monoxide and hydrogen an aldehyde is produced in good yield.

$$RCH{=}CH_2 + CO + H_2 \rightarrow RCH_2CH_2CHO$$

This can be followed by reduction to a primary alcohol, and is an important industrial process for obtaining primary alcohols.

5 *Oxidation* – The oxidation of alkenes can lead to a variety of products of which the epoxides such as ethylene oxide are of particular importance industrially. These compounds are of use as chemical intermediates; ethylene oxide for example is used to make ethane-1,2-diol (glycol) used as an antifreeze.

$$CH_2{=}CH_2 + O_2 \xrightarrow[\text{catalyst}]{\text{silver}} \overset{\;\;O}{CH_2{-}CH_2} \xrightarrow{H_2O} CH_2OHCH_2OH$$

6 *Addition of halogen* – The chlorination and bromination of ethylene are industrially important. Both reactions occur readily at ordinary pressure and temperature.

$$CH_2{=}CH_2 + Cl_2 \rightarrow CH_2ClCH_2Cl$$
$$CH_2{=}CH_2 + Br_2 \rightarrow CH_2BrCH_2Br$$

7 *Polymerization* – Polythene is made by polymerizing ethylene. The polymerization is carried out either by a high pressure process, or at lower pressures in the presence of a catalyst. In the ICI high pressure process, discovered in 1933, ethylene is polymerized at a pressure of above 1000 atmospheres and at about 200 °C in the presence of a trace of oxygen. In the Ziegler process, developed during the 1950s, ethylene is dissolved in a hydrocarbon solvent in which is suspended a catalyst of an aluminium trialkyl and titanium tetrachloride. The temperature used is about 70 °C and the pressure about 10 atmospheres.

$$n\,CH_2{=}CH_2 \rightarrow ({-}CH_2CH_2{-})_n$$

The reactions of alkenes are obviously addition reactions. How many of the reactions listed could occur by a mechanism similar to that discovered by Francis for the room temperature bromination of ethylene (section 9.2)?

Revision questions

You should now be able to deduce what reagents, organic and inorganic, are required for the synthesis in one step of the following products from the appropriate alkene:

1 2-bromo-octane
2 pentanal
3 butan-2-ol
4 butan-1-ol (in two steps).

13.2 Unsaturated compounds: the carbonyl group in aldehydes and ketones

This section is concerned with two closely related types of compound, the aldehydes and the ketones. Both of these types contain the carbonyl or C=O group in their molecular structure, and so they have many properties in common.

Aldehydes have a structure in which one alkyl group and one hydrogen atom are attached to the carbonyl group. Their formula can therefore be written

```
R
 \
  C=O
 /
H
```

where R stands for an alkyl group. This is usually written RCHO. If a CH_3 group is attached to the functional group, the structure CH_3CHO is obtained; the compound having this structure is known as acetaldehyde (or ethanal). A space-filling model of this molecule is shown in figure 13.2.

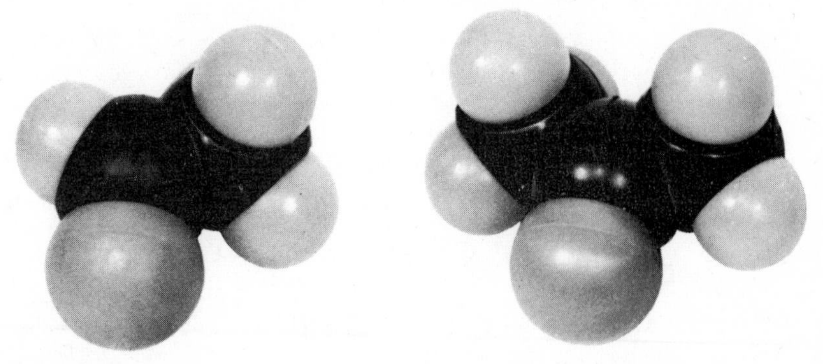

Figure 13.2
Space-filling models representing an aldehyde and a ketone (acetaldehyde and acetone).

Ketones have a structure in which two alkyl groups are attached to the carbonyl group. Their formula can be written

$$\begin{array}{l} R \\ \;\; \backslash \\ \;\;\; C{=}O \\ \;\; / \\ R \end{array}$$

or alternatively as RCOR. When two methyl groups are attached to the carbonyl group, CH_3COCH_3, the structure obtained is that of acetone (or propanone). A space-filling model of this molecule is also shown in figure 13.2.

For compounds as closely related as the aldehydes and ketones, with many properties in common, some care must be taken in distinguishing between them. The purpose of the next experiment is to show how this can be done. In the first part you will find out how to distinguish between aldehydes and ketones using a simple test, and in the second part how to identify a member of either of these series.

Identification of individual members of a series, such as the series of aldehydes, is usually done by preparing a crystalline derivative and taking its melting point. For carbonyl compounds the best derivative to make is that formed by reaction with 2,4-dinitrophenylhydrazine.

$$(CH_3)_2C{=}O + NH_2{-}NH{-}C_6H_3(NO_2)_2$$

acetone and 2,4-dinitrophenylhydrazine

$$\downarrow$$

$$\left[(CH_3)_2C(OH){-}N(H){-}NH{-}C_6H_3(NO_2)_2 \right]$$

$$\downarrow$$

$$(CH_3)_2C{=}N{-}NH{-}C_6H_3(NO_2)_2 + H_2O$$

acetone 2,4-dinitrophenylhydrazone

The reasons for using such a complicated reagent are as follows

When preparing a derivative for identification purposes, the compound chosen should be capable of reacting with *all* the members of the particular homologous series, to give a derivative having the following properties.

1 It must be easily prepared by a reaction of high yield, involving little or no formation of by-products.

2 It must be a stable compound not decomposed at temperatures below or at its melting point.

3 It must be easily purified by recrystallization.

4 It should have a melting point in the range 50–250 °C, for convenience in determining the melting point.

These conditions rule out a number of possible compounds. In order to get melting points in a suitable range, compounds of fairly large molecular weight have to be made. As an example of this, the products formed by reactions between carbonyl compounds and hydrazine, NH_2NH_2, and its derivatives may be quoted. Hydrazine itself does not form products which are crystalline solids, but its derivative, phenylhydrazine, does in many cases. Increasing the molecular weight still more by the introduction of the two nitro-groups to give 2,4-dinitrophenylhydrazine ensures the formation of a crystalline product.

Experiment 13.2

Identification of an aldehyde or ketone

In this experiment you will try out one reaction which can be used to distinguish between aldehydes and ketones. You will then use this to test an unknown substance belonging to one of these two classes of compounds. Finally you will prepare a derivative of the unknown, and try to identify it from its melting point.

Procedure

1 You are provided with a substance known to be either an aldehyde or a ketone. Try out the following test to decide which it is. Use a 150 × 25 mm test-tube for this test. Boil a little Fehling's solution (a mild oxidizing agent) with an aldehyde for a few minutes. Now boil some with a ketone. What difference do you notice? Boil a little of the unknown substance with Fehling's solution, and decide to which group it belongs.

2 Now take a few drops of the unknown aldehyde or ketone and dissolve them in the *minimum* quantity of methanol. Add this solution to about 5 cm^3 of the solution of 2,4-dinitrophenylhydrazine, shake the mixture, and allow it to stand. If no precipitate appears, carefully add 1–2 cm^3 of dilute sulphuric acid. Filter off the yellow precipitate using suction filtration, wash it (with the suction disconnected) with 1 cm^3 of methanol, and dry it by sucking air through it for a few minutes.

The solid obtained should now be crystallized by dissolving it in the minimum quantity possible of a hot 1:1 ethanol-water mixture and allowing the solution to cool. The crystals obtained should then be filtered and dried as before, and their melting point obtained. The procedure for finding melting points was described in section 9.4.

Compare the melting point of your crystals with the values given in the table below and see if you can identify the material. Now check the boiling point of the unknown substance (see below) and compare with those in table 13.2.

Name	Formula	Boiling point/°C	Melting point of 2,4-dinitrophenyl-hydrazone/°C
Aldehydes			
Formaldehyde (methanal)	HCHO	–19	167
Acetaldehyde (ethanal)	CH_3CHO	21	164, 146 (2 forms)
Propanal	CH_3CH_2CHO	48	156
Butanal	$CH_3CH_2CH_2CHO$	75	123
2-Methylpropanal	$(CH_3)_2CHCHO$	64	187
Pentanal	$CH_3CH_2CH_2CH_2CHO$	104	98
Benzaldehyde	C_6H_5—CHO	178	237
Ketones			
Acetone (propanone)	CH_3COCH_3	56	128
Butanone	$CH_3CH_2COCH_3$	80	115
Pentan-2-one	$CH_3CH_2CH_2COCH_3$	102	141
Pentan-3-one	$CH_3CH_2COCH_2CH_3$	102	156
3-Methylbutanone	$(CH_3)_2CHCOCH_3$	94	117
Hexan-2-one	$CH_3CH_2CH_2CH_2COCH_3$	128	107
Cyclohexanone	C_6H_{10}=O	156	162
Acetophenone	C_6H_5—$COCH_3$	202	250, 237 (2 forms)

Table 13.2
Physical data for some aldehydes and ketones and their 2,4-dinitrophenylhydrazones

3 Boiling points are most easily found by gently boiling 0.5 to 1.0 cm^3 of the liquid, in contact with one or two anti-bumping granules, in a 150 × 16 mm test-tube, with a thermometer suspended just above the level of the liquid. Best results are obtained if the test-tube is immersed in a beaker containing dibutylphthalate. This should be heated until the thermometer shows a constant reading, and the ring of liquid refluxing in the test-tube is about 1 cm above the bulb of the thermometer.

State what you consider your unknown compound to be.

Summary of reactions of carbonyl compounds

1 *Reduction with hydrogen* – Hydrogen at ordinary temperature and pressure will react with carbonyl compounds if a nickel catalyst is used, producing alcohols in excellent yield. Aldehydes give primary alcohols and ketones give secondary alcohols.

$$RCHO + H_2 \rightarrow RCH_2OH$$
$$RCOR + H_2 \rightarrow RCHOHR$$

2 *Reaction with hydrogen cyanide* – Carbonyl compounds react with hydrogen cyanide, in the presence of a little alkali to provide some CN^- ions, to give cyanohydrins. The reaction is useful because the cyanide group enables a range of bifunctional compounds to be prepared e.g. α- hydroxycarboxylic acids.

$$\mathrm{R(H)C{=}O} + HCN \rightarrow \mathrm{R(H)C(OH)(CN)} \xrightarrow{H_2O} RCHOHCO_2H$$

If ammonium cyanide ($NH_4Cl + NaCN$) is used, the product is an α-amino carboxylic acid.

$$\mathrm{R(H)C{=}O} + NH_4^+CN^- \rightarrow \mathrm{R(H)C(NH_2)(CN)} \xrightarrow{H_2O} RCHNH_2CO_2H$$

3 *Reaction with ammonia* – With ammonia alone the reactions of carbonyl compounds are complex because a variable sequence of reactions can occur. If, however, carbonyl compounds are treated with hydrogen in the presence of excess alcoholic ammonia, using a nickel catalyst, primary amines can be obtained in yields of 50 per cent or better.

$$\mathrm{R(H)C{=}O} + NH_3 + H_2 \rightarrow RCH_2NH_2 + H_2O$$

4 *Reaction with derivatives of ammonia* – Hydroxylamine, NH_2OH, adds to carbonyl compounds to give oximes. The oxime can then be reduced by hydrogen, from sodium and alcohol, or hydrogen under pressure, with a nickel catalyst, to produce amines in yields up to 90 per cent.

$$\text{RHC=O} + NH_2OH \rightarrow \text{RHC(OH)(NHOH)} \rightarrow \text{RHC=NOH} \xrightarrow{H_2} RCH_2NH_2 + H_2O$$

Reaction also occurs with derivatives of hydrazine, including phenylhydrazine, C_6H_5—$NHNH_2$ and 2,4-dinitrophenylhydrazine, NO_2—$C_6H_3(NO_2)$—$NHNH_2$.

The products are readily purified, have sharp melting points, and are therefore used to identify carbonyl compounds (see experiment 13.2).

5 *Reaction with halogens and hydrogen halides* – Bromine reacts with carbonyl compounds dissolved in warm aqueous acetic acid to give monosubstitution products in good yield.

$$CH_3COCH_2CH_3 + Br_2 \xrightarrow{50\%} CH_3COCHBrCH_3 + HBr$$

$$CH_3COCH_2CH_3 + Br_2 \xrightarrow{20\%} CH_2BrCOCH_2CH_3 + HBr$$

Addition of hydrogen halides to the carbonyl double bond does occur but the reaction readily reverses and the products cannot usually be isolated.

6 *Oxidation* – Dichromates and other mild oxidizing agents convert aldehydes to carboxylic acids; ketones cannot be oxidized without splitting a carbon-carbon bond.

$$3RCHO + Cr_2O_7^{2-} + 8H^+ \rightarrow 3RCO_2H + 2Cr^{3+} + 4H_2O$$

You should inspect these reactions of the carbonyl group, noting the possible polarization of the C=O bond and hence the probable nature of the attacking agent. What similarities and differences do you note with addition to alkenes?

Revision questions

You should now be able to deduce what reagents, organic and inorganic, are required for the synthesis in one step of the following products:

1 dodecylamine
2 butan-2-ol
3 pentanoic acid
4 1-hydroxyoctanoic acid (two steps).

Background reading

Diabetes

Sufferers from diabetes are always liable to lapse into a coma if their regular treatment is neglected. Early symptoms of the onset of a coma include thirst and drowsiness, and tests on the patient's urine will reveal the presence of glucose and acetone. If no treatment is undertaken coma and eventual death may occur, these events being paralleled by an increase of acetone in the urine; acetone may even be smelt in the breath.

Acetone, and other keto compounds, are produced by the metabolism of fats and this is a normal process. But in the absence of other energy sources the metabolism of fats occurs to excess and the quantity of keto compounds in the blood rises to a toxic concentration.

In normal human metabolism glucose is converted to a polymer, glycogen, in the liver and kept there and in muscles as an energy store, being broken down to glucose again as required. The availability of glycogen and the ability of the muscles to utilize glucose are normally enough to prevent excessive metabolism of fats. But in people suffering from diabetes the ability of the liver to convert glucose to glycogen and of muscles to utilize glucose has largely been lost. Hence, if the condition is untreated, the concentration of glucose in the blood stream will rise, and eventually glucose will be lost in the urine. If the condition remains untreated an increasing amount of fat will be used as an energy source and the products of metabolism, including acetone, will also appear in the urine.

It has been found that in diabetes the pancreas gland is not producing sufficient of a complex protein called insulin, which appears to control glucose metabolism. Regular treatment with insulin and some care with the diet will maintain a properly balanced metabolism and prevent the production of toxic concentrations of acetone. The structure of insulin is discussed in Topic 18.

To test for reducing sugars (glucose) in urine

Add 8 drops of urine to 5 cm^3 of Benedict's solution and boil vigorously for two minutes. If more than 0.2 g of sugar per 100 cm^3 of urine are present a yellow to brown deposit of copper(I) oxide will be produced and the solution will become colourless; if less sugar is present the solution will become green and only a small deposit will form.

To test for ketones (acetone) in urine

To 2 cm^3 of urine add a spatula full of mixed powdered ammonium sulphate and sodium nitroprusside (100/1). Add 1 cm^3 of 0.880 ammonia. A faint purple colour will develop if acetone is present in a proportion of 1 in 20000 or more.

13.3 Unsaturated compounds: the arenes

The structure of benzene, C_6H_6, provided chemists with a major problem, the principal difficulty being the existence of only one monosubstituted derivative of benzene, C_6H_5X. An acceptable structure had therefore to fulfil the condition that all six hydrogen atoms must occupy equivalent positions.

A major step towards the solution to the problem was taken by Kekulé, then Professor of Chemistry at Ghent, in 1865. He later described how he came to propose the structure illustrated in figure 13.3b(i).

> I turned my chair to the fire and dozed [he relates]. Again the atoms were gambolling before my eyes. This time the smaller groups kept modestly in the background. My mental eye, rendered more acute by repeated visions of this kind, could now distinguish larger structures, of manifold conformation; long rows, sometimes more closely fitted together; all twining and twisting in snakelike motion. But look! What was that? One of the snakes had seized hold of its own tail, and the form whirled mockingly before my eyes. As if by a flash of lightning I awoke.

Arthur Koestler, in his book *The Act of Creation*, from which this translation is taken, describes this as probably the most important dream in history since Joseph's seven fat and seven lean cows. 'The serpent biting its own tail', he writes, 'gave Kekulé the clue to a discovery which has been called "the most brilliant piece of prediction to be found in the whole range of organic chemistry" and which, in fact, is one of the cornerstones of modern science.' It was the first suggestion that carbon atoms in molecules did not only form chains, but also rings, like the snake swallowing its tail.

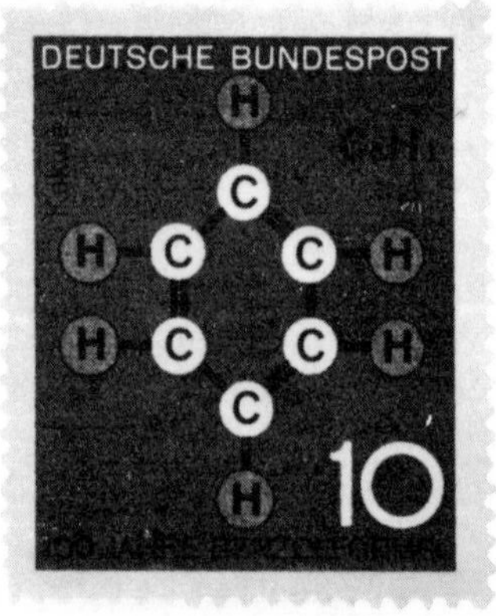

Figure 13.3a
Stamps issued to commemorate Kekulé's proposal of a ring structure for benzene.

The centenary of this event was commemorated by, amongst other things, the issue of postage stamps (figure 13.3a).

The modern evidence for the symmetry of the benzene ring is based on X-ray diffraction studies. The unusual nature of the bonding is seen from a comparison of the bond lengths of benzene with those of cyclohexene.

carbon-carbon single bond in cyclohexane	0.154 nm
carbon-carbon double bond in cyclohexene	0.133 nm
carbon-carbon all bonds in benzene	0.139 nm

When drawing a structure to indicate the molecule of benzene certain representational difficulties arise; a single line is normally used to represent two electrons, and two lines to represent four electrons. As neither of these is appropriate for the carbon-carbon bonds in benzene the representation given in figure 13.3b(ii) is used.

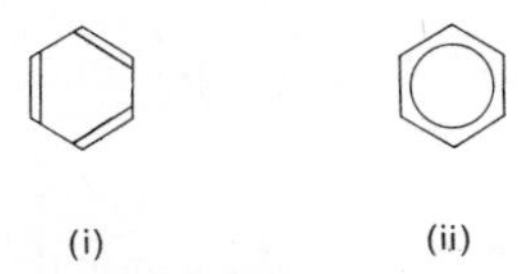

Figure 13.3b
Kekulé and contemporary representations of benzene.

This delocalization of electrons has already been discussed in section 8.4 and its effect on the chemical reactions of benzene noted in section 9.2. This effect may be summarized by saying that whereas cyclohexene undergoes reactions, for example with sulphuric acid, that are mostly *addition reactions*,

$$C_6H_{10} + H_2SO_4 \rightarrow C_6H_{10}(H)(HSO_4)$$

benzene (which from the Kekulé structure might be thought to be more prone to additions) in similar circumstances undergoes *substitution reactions*

$$C_6H_6 + H_2SO_4 \rightarrow C_6H_5SO_3H + H_2O$$

In Topic 9 the influence of the benzene ring on the behaviour of functional groups was examined. In this section the substitution reactions of benzene and the influence of those same functional groups on the benzene ring are to be investigated.

Before you begin this work make sure you understand the structure of benzene and also the other arenes to be considered, phenol and chlorobenzene (figure 13.3c).

Figure 13.3c
Space-filling models representing benzene, phenol, and chlorobenzene.

Experiment 13.3a

The bromination of arenes

Caution. Remember phenol is corrosive and benzene vapour is poisonous.

Procedure

1 To about 5 cm^3 of bromine water in a test-tube add a few drops of benzene. Shake well and observe what happens to the colour of the bromine.

Benzene is a good solvent for bromine but you may remember (from section 9.2) that it contrasts with alkenes as it does not react readily with bromine to give a colourless product.

2 Now repeat the first experiment but this time add a few crystals of phenol to some bromine water, and shake well. What has happened to the colour of the bromine now? What does this suggest? What is the appearance of the mixture after five minutes?

The crystalline material that you should now be able to see is 2,4,6-tribromophenol. The equation for its formation is

OH (on benzene ring) $+ 3Br_2 \rightarrow$ OH, Br, Br, Br (on benzene ring) $+ 3HBr$

This type of reaction, in which a hydrogen atom of a benzene ring is replaced by another atom, is known as a substitution reaction. You have seen that it takes place quite easily with phenol but not with benzene. Benzene will undergo such a reaction but it requires more drastic conditions; bromine (not bromine water) is needed, and a catalyst (iron, iodine, and some other materials are suitable) must be present.

$$C_6H_6 + Br_2 \xrightarrow{\text{catalyst}} C_6H_5Br + HBr$$

Even then, the major product is the monobromo compound, not the tribromo compound as with phenol, although the yields of dibromo- and tribromobenzene can be increased by heating.

It therefore appears that the hydroxyl group has an effect on the benzene ring in this case.

3 You may like to examine the bromination of benzene, but this operation *must be done in a fume cupboard*. Put about 2 cm^3 of benzene in a test-tube and using a teat pipette add 2–3 drops of bromine. (*Caution*. Exercise *great* care when dealing with bromine and do not spill any on your skin.) Does a reaction take place?

Now add some iron (a tin tack is suitable).
Is a reaction taking place now? What fumes are evolved?

4 In this experiment the bromination of chlorobenzene will be attempted. To about 5 cm^3 of bromine water in a test-tube add a few drops of chlorobenzene. Shake well and observe what happens to the colour of the bromine.

Is there any sign of reaction? Remember that chlorobenzene is likely to be a good solvent for bromine just as benzene was in part 1.

Now try the more drastic conditions of part 3. Put about 2 cm^3 of chlorobenzene in a test-tube and then, in a fume cupboard, add 2–3 drops of bromine using a teat pipette (*caution*). Does a reaction take place? Add some iron (a tin tack) as a catalyst.

Is a reaction taking place now? What fumes, if any, are evolved?

The substitution reaction of arenes

We can now consider how the bromine substitution reaction of the benzene ring takes place.

What evidence is there of the nature of the attacking group? L. J. Lambourne and P. W. Robertson (*J. Chem. Soc.*, p. 1167, 1947) found that the reaction of iodine monochloride, I—Cl, with arenes produced only iodine substitution products.

$$C_6H_5OCH_3 + I{-}Cl \rightarrow p\text{-}IC_6H_4OCH_3 + HCl$$

What is the attacking atom in this reaction?

What polarization would you expect in I—Cl? So what is the charge on the attacking atom in the reaction?

Now consider the leaving group. What atom is lost from the benzene ring in the reaction? Would this atom more easily carry a positive or a negative charge when it leaves the benzene ring? Is this consistent with the charge you have proposed that the attacking atom will bring to the benzene ring?

You should now have a hypothesis about the charge on the attacking agent in an arene substitution and also about the nature and charge of the leaving group. We can now see if this hypothesis is consistent with the relative ease of attack on benzene, phenol, and chlorobenzene.

What polarization of the benzene ring is required to facilitate attack by the iodine atom of iodine monochloride? Dipole moments can be quoted as suggestive of the polarization of the benzene ring caused by substituents.

Molecule	Dipole moment $(\delta^-) \leftrightarrow +$
$C_6H_5{-}OH$	1.6
C_6H_6	0.0
$Cl{-}C_6H_5$	1.6

Table 13.3a
Dipole moments of phenol, benzene, and chlorobenzene

In the case of phenol, is the dipole moment consistent with the charge required on the benzene ring to activate the substitution reaction? In the case of chlorobenzene is the dipole moment consistent with the charge required to deactivate the reaction?

Finally, let us examine the function of the iron catalyst.

Iron reacts with bromine to form iron(III) bromide

$$2Fe + 3Br_2 \rightarrow 2FeBr_3$$

This in turn induces polarization in other bromine molecules,

$$FeBr_3 + Br_2 \rightarrow Br^{\delta+}—Br^{\delta-}.\ FeBr_3$$

Reaction of this last compound with the arene regenerates the iron(III) bromide, and the catalyst is therefore iron(III) bromide and not iron.

Thus the function of the catalyst is to provide a bromine atom carrying the correct charge for attack on the benzene ring.

Experiment 13.3b

The nitration of arenes

Caution. You will be using a mixture of concentrated nitric and sulphuric acids which is dangerous, and great care must be taken to avoid getting any on your hands, clothes, or elsewhere. If any is accidentally spilt, a large quantity of water should be poured on it at once.

Procedure

1 The first experiment is the attempted nitration of benzene in mild conditions.

To about 6 cm^3 of dilute sulphuric acid in a test-tube add 2 g of sodium nitrate and shake well to dissolve. Add about 1 cm^3 of benzene, shake well, and observe what happens.

2 The second experiment is the attempted nitration of benzene in more drastic conditions.

By means of a teat pipette, carefully add, drop by drop, about 0.5 cm^3 of benzene to a mixture of 1 cm^3 each of concentrated nitric and sulphuric acids. Shake the mixture well after the addition of each drop but be careful not to let any of the mixed acids splash out of the test-tube. The mixture will get hot. When the change appears to be complete, pour the contents of the test-tube into a small beaker two-thirds full of water, and stir well.

Does your result suggest that a reaction has occurred? Use the information in table 13.3b on the appearance and density of possible products to help you decide.

Substance	Appearance	Density /g cm^{-3}	Melting point/°C
Benzene	liquid	0.88	5.7
Nitrobenzene	pale yellow liquid	1.2	5.9
m-Dinitrobenzene	yellow solid	—	90

Table 13.3b
Possible products in the attempted nitration of benzene

3 In this experiment the nitration of phenol will be attempted using the mild conditions of part 1 above.

To about 6 cm^3 of dilute sulphuric acid in a test-tube add 2 g of sodium nitrate and shake well to dissolve. Cool the solution by standing the test-tube in an ice bath.

Melt 1 g of phenol with about 1 cm^3 of water in another test-tube and add this mixture to the cooled nitrate solution, a portion at a time, so that the temperature does not exceed 20 °C. Shake well after each addition of phenol. When all the phenol has been added, allow the reaction mixture to stand for one hour, shaking it frequently. (Portions may however be removed for running chromatograms before an hour is complete.)

Is there a difference between the behaviour of phenol and that of benzene?

Decant the excess of reaction solution from the dark resinous product and wash the product two or three times with 10 cm^3 portions of water, discarding the washings. Transfer the liquid product to a watchglass where the last droplets of water should be removed by strips of filter paper. The watchglass should be kept covered to prevent loss of any volatile material. If the product has solidified, it should be dried on a porous plate. The material that has been made is a mixture of nitrophenols and other products. The technique of thin-layer chromatography can be used to separate them and to attempt to identify some of them if you have sufficient time.

Silica gel is used for preparing the thin-layer and the chromatography is carried out with trichloromethane as solvent.

Thin-layer chromatography of the products

Two of the products of this reaction are *ortho*- and *para*-nitrophenol. Their presence can be demonstrated by running samples of the authentic materials by the side of the reaction product mixture.

Chromatography plates 10 × 5 cm are required. The silica gel layer should be scraped off at the edges of the plate to prevent the solvent from running unevenly.

Melting point tubes drawn out to a fine jet are used for spotting the plates with samples of *ortho*- and *para*-nitrophenol and the reaction product. Solutions of a few small crystals of the nitrophenols may be made in ethanol in test-tubes which should be kept corked. A portion of the nitrated product is also dissolved in ethanol and kept in a corked test-tube.

One drop of each of the three samples is placed carefully in line about 1 cm from a narrow end of the plate, at equal distances from each other. The drops should be at least 1 cm from the edges of the plate and not exceed 2–3 mm in diameter when dried (figure 13.3d). A common error is to have too big a drop.

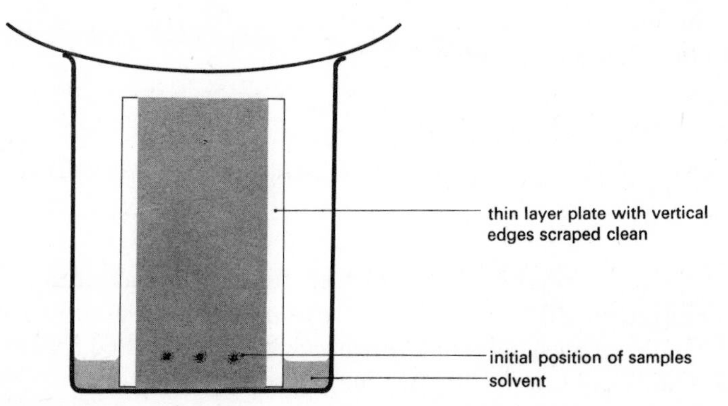

Figure 13.3d
Apparatus for thin-layer chromatography.

The plate is now placed in a beaker containing about 5 mm depth of trichloromethane, and covered. The plate should stand as near vertical as possible, with the spots along the bottom edge nearest to the trichloromethane. Separation of coloured bands for the reaction product occurs after about 10 minutes. The separations should be appreciable, and can be matched against the yellow colours of the *ortho*- and *para*-nitrophenol samples used, when the plate is removed from the trichloromethane and has dried.

How many products are made by this reaction? Do you think that you can identify any of them as either *ortho*- or *para*-nitrophenol?

4 In this experiment the nitration of chlorobenzene will be attempted using the more drastic conditions of part 2 above.

By means of a teat pipette, carefully add, drop by drop, about 0.5 cm^3 of chlorobenzene to a mixture of 1 cm^3 each of concentrated nitric and sulphuric acids. Shake the mixture well after the addition of each drop but be careful not to let any of the mixed acids splash out of the test-tube. Look for evidence of a reaction taking place, such as increase in temperature, and the evolution of nitrogen dioxide. Finally pour the product into water. If no reaction has taken place, no yellow oil will separate (table 13.3c).

Substance	**Appearance**	**Density /g cm^{-3}**	**Melting point/°C**
Chlorobenzene	liquid	1.1	—
o-Chloronitrobenzene	yellow liquid	1.4	—
p-Chloronitrobenzene	yellow solid	—	83
1-Chloro-2,4-dinitrobenzene	yellow solid	—	52

Table 13.3c
Possible products from chlorobenzene nitration

Does the chlorine atom have any noticeable effect on the ease of nitration of the benzene ring?

5 If, in the conditions of experiment 4, chlorobenzene was nitrated readily, see if it will nitrate in milder conditions, for example by the addition of dilute nitric acid. On the other hand if chlorobenzene was not nitrated try even more drastic conditions, for example by heating the reaction mixture.

Draw up a table summarizing your results in these experiments comparing the reactivity of benzene, chlorobenzene, and phenol on the basis of the experimental work you have carried out on nitration and bromination.

You will be discussing in class the process of nitration of arenes to see if it is comparable to the process of bromination.

Background reading

The manufacture of phenol

The history of the manufacture of phenol provides an interesting example of how social and financial considerations influence the choice of chemical reactions used in industrial processes.

The original large-scale source of phenol was coal tar, which contains some two hundred organic compounds. Phenol occurs in the 'middle oil' fraction of coal tar. This is obtained by fractional distillation of coal tar, collecting the fraction which boils over the range 170–230 °C. Phenol can be isolated from this in the following steps.

All the acidic components of the middle oil fraction are extracted by sodium hydroxide and then the phenols are recovered from the extract by treatment with carbon dioxide. Carbon dioxide converts sodium hydroxide, a strong base, to sodium carbonate, a weak base, in which phenols are much less soluble. The phenols are then separated by fractional distillation, phenol itself boiling at 182 °C, the cresols from 190 to 203 °C, and the xylenols from 211 to 225 °C.

The First World War brought such a sudden large increase in the demand for phenol that the supplies from coal tar were no longer sufficient. Phenol was required for conversion to picric acid, which was used as an explosive, and to cope with this sudden increased demand the first synthetic method for phenol manufacture was introduced.

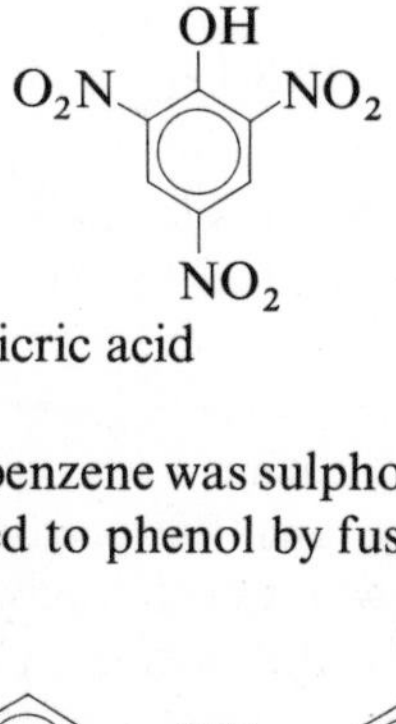

OH
O_2N NO_2
NO_2
picric acid

In this, benzene was sulphonated, and the benzenesulphonic acid thus produced converted to phenol by fusing with alkali.

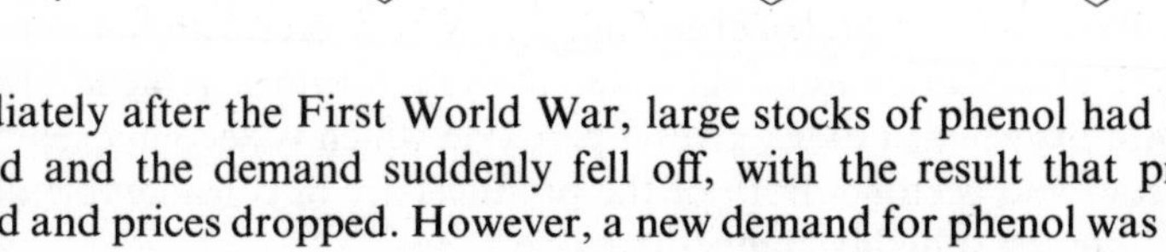

$$C_6H_6 \xrightarrow{H_2SO_4} C_6H_5SO_3H \xrightarrow{Na^+OH^-} C_6H_5O^-Na^+ \xrightarrow{H^+} C_6H_5OH$$

Immediately after the First World War, large stocks of phenol had been synthesized and the demand suddenly fell off, with the result that production stopped and prices dropped. However, a new demand for phenol was shortly to come from a new industry, the manufacture of plastics, using phenol-formaldehyde polymers. The stock remaining from war production was rapidly consumed, but, as this had been sold at less than cost price, a cheaper and more efficient method of manufacture was wanted to prevent a severe rise in prices.

This resulted in the discovery of a new process by the Dow Chemical Company in America. In this, chlorobenzene is heated with aqueous sodium hydroxide under pressure at 300 °C.

$$C_6H_6 \xrightarrow{Cl_2} C_6H_5Cl \xrightarrow{Na^+OH^-} C_6H_5O^-Na^+ + NaCl$$

The Dow process had the disadvantage of using up chlorine, which has to be manufactured, and yielding as its only by-product sodium chloride, a common naturally occurring substance.

These disadvantages were removed in the Raschig process, developed in Germany. This is a two-stage process. In the first stage benzene, hydrogen chloride, and air are passed over a catalyst of copper chloride at 200 °C, to produce chlorobenzene and water. In the second stage the chlorobenzene and steam are passed over a second catalyst of silicic acid at 400 °C, and react to give phenol and hydrogen chloride.

$$C_6H_6 \xrightarrow{HCl + O_2(air)} C_6H_5Cl \xrightarrow{H_2O} C_6H_5OH + HCl$$

It can be seen that in the overall reaction the only substances actually used up are benzene and air, because the hydrogen chloride used in the first stage is regenerated in the second.

During and after the Second World War petroleum became a very important source of raw materials for the chemical industry, and the petrochemical industry has grown to be an industry in its own right. A process for manufacturing phenol from petroleum products was developed by the Distillers Company in Great Britain.

Propylene, from the cracking of crude oil, is made to react with hot benzene in the presence of an aluminium chloride catalyst, forming cumene. This, on reaction with oxygen at 110 °C, gives a peroxide which is decomposed by acid to give phenol and acetone. Both of the products are in considerable demand making the process a highly competitive one.

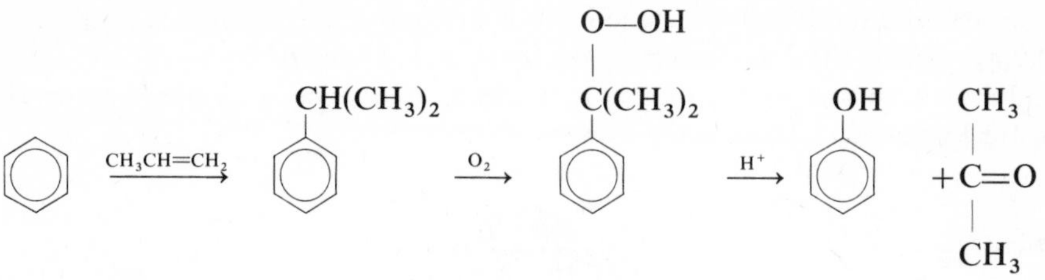

$$C_6H_6 \xrightarrow{CH_3CH{=}CH_2} C_6H_5CH(CH_3)_2 \xrightarrow{O_2} C_6H_5C(CH_3)_2\text{—O—OH} \xrightarrow{H^+} C_6H_5OH + (CH_3)_2C{=}O$$

Summary of the reactions of arenes

1 *Nitration* – Nitric acid, usually in the presence of concentrated sulphuric acid to produce nitronium ions NO_2^+, substitutes a nitro group —NO_2 for a hydrogen atom.

$$C_6H_6 + NO_2^+ \rightarrow C_6H_5NO_2 + H^+$$

2 *Halogenation* – Bromine, in the presence of catalysts such as iron(III) bromide, substitutes a bromine atom for a hydrogen atom.

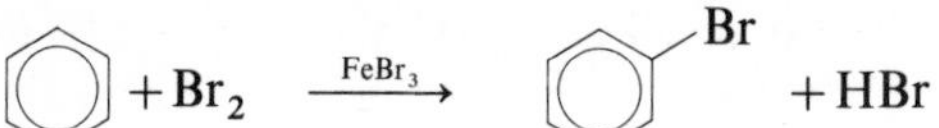

$$C_6H_6 + Br_2 \xrightarrow{FeBr_3} C_6H_5Br + HBr$$

3 *Sulphonation* – Fuming sulphuric acid is used for sulphonation. The reactive entity is considered to be $^{\delta+}SO_3^{\delta-}$

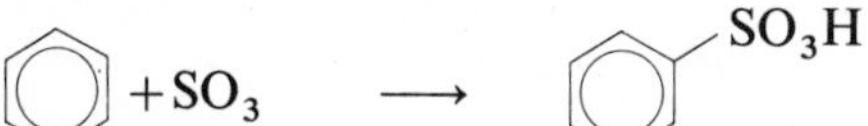

$$C_6H_6 + SO_3 \longrightarrow C_6H_5SO_3H$$

Although other reactions, including addition in drastic conditions (for example the preparation of BHC, see Topic 9) are possible for arenes, the three listed above the most important. All the above reactions involve attack by cations or positively polarized atoms resulting in substitution for hydrogen which leaves as a proton.

13.4 Amines

In this section you will investigate some carbon compounds which can be regarded as derivatives of ammonia, as their structures are obtained by replacing a hydrogen atom in the ammonia molecule with a hydrocarbon group. The resulting compound is called a *primary amine*.

You will be looking at the reactions of ammonia, butylamine, and aniline. Before starting the experiments, make sure you are familiar with their formulae and molecular shapes, if possible by building models such as those shown in figure 13.4a.

Figure 13.4a
Space-filling models representing butylamine, aniline, and acetamide.

Experiment 13.4

What is the effect of putting an alkyl or aryl group into the ammonia molecule?

Procedure

1 To ammonia solution, and drops of butylamine, and aniline, separately shaken with water, add a drop of universal indicator. Are the amines soluble in water? What type of molecular interaction will be helping them dissolve? Are the amines basic? Can you estimate their relative basic strength?

2 Add drops of the amines to dilute hydrochloric acid.
Are the amines more soluble in hydrochloric acid than in water?
Could they be reacting with the hydrochloric acid?
Write down equations representing any reactions which you consider to be taking place.
How would you recover the butylamine or aniline from their mixture with dilute hydrochloric acid? Test your suggestion experimentally.

3 Add separately drops of the amines and ammonia to copper sulphate solution until present in excess. Are somewhat similar results obtained?
What type of interaction do you imagine is taking place?

4 Is ammonia gas inflammable? Do you suppose that the amines might be inflammable? Test your suppositions using a few drops of each of them.

5 How does ammonia react with nitrous acid? To a spatula full of ammonium chloride dissolved in a little hot water add an equal amount of solid sodium nitrite.
What is the unreactive gas evolved?

6 The reaction of the amines with nitrous acid can now be investigated, but at a lower temperature. Three reaction mixtures are to be prepared. Butylamine will be used as an example of an alkyl amine and aniline as an example of an aryl amine. The third mixture is a 'blank' for comparison.
To about 25 cm^3 of a crushed ice–water mixture at 5–10 °C in a 250 cm^3 beaker, add 0.5 cm^3 of butylamine previously dissolved in 10 cm^3 of 2M hydrochloric acid. Now add in small portions, stirring well, a solution of 0.5 g sodium nitrite in 10 cm^3 of water.
Prepare a similar reaction mixture using 0.5 cm^3 of aniline in place of the butylamine.
Prepare a 'blank' reaction mixture using the inorganic reagents only and omitting any amine.
Allow the three reaction mixtures to stand for five minutes (but no longer) and in the meantime prepare a solution of 3 g of 2-naphthol in 20 cm^3 of 2M sodium hydroxide. Warm if necessary to dissolve, divide into three equal portions, and dilute each with 50 cm^3 of cool water. At the end of five minutes add small portions of the three reaction mixtures to the separate portions of 2-naphthol solution. Does it look as if the amines have given distinctive reaction products? Finally add all of the reaction mixtures to the 2-naphthol solutions.
What does the 'blank' mixture tell you?

Aryl amines played an important part in the development of the organic chemical industry, as will be seen from the account which now follows.

Background reading

Dyestuffs: the origins of the modern organic chemical industry

Edmund Burke said, 'People will not look forward to posterity, who never look backward to their ancestors.' We should certainly not neglect the tremendous contributions to the present state of our science which were made by those individuals who effectively laid the foundations of the modern organic chemical industry.

In the years before 1850 the organic chemical industry scarcely existed, and nothing in the progress of industrial chemistry has been more spectacular than its emergence, involving as it does the manufacture of thousands of complex substances, including dyes, drugs, explosives, plastics, man-made fibres, fuels, plant protection chemicals, insecticides, and a host of others.

There is a marked difference in character between the inorganic and organic industrial scenes. In the former, developed primarily in Britain and France during the first half of the nineteenth century, the chemist devised processes for the manufacture of heavy chemicals such as iron, steel, sulphuric acid, caustic soda, and ammonia. These processes, once developed, could then be carried on by trained workers for many years, the chemist himself being involved mainly in a trouble-shooting role. The organic chemical industry, however, is one that changes so rapidly in character, with the frequent discovery of new compounds and new synthetic routes, that the chemist is continuously involved. The methods of the organic industry were laid down during the establishment of the synthetic dyestuff and drug industries in the latter half of the nineteenth century.

Let us consider some aspects of the story of the dyestuff industry because the principles of working which governed it are still basic to the philosophy of our modern organic chemical industry. We must begin with the discovery, in 1856, by an eighteen-year-old student, W. H. Perkin, of the first synthetic colouring material, a purple dye known as aniline purple or mauve. Perkin was a student of the German chemist, Hofmann, who was the Professor of Chemistry at the Royal College of Chemistry in London. Hofmann suggested that the drug quinine might be synthesized from aromatic amines derived from coal tar, and Perkin, on his own initiative, set out to attempt this. At that time the structural formulae of organic compounds had not been worked out, and chemists knew only the molecular formulae, that for quinine being $C_{20}H_{24}N_2O_2$. Starting with the amine allyltoluidine, whose empirical formula is $C_{10}H_{13}N$, Perkin attempted his synthesis on the basis of the proposed reaction,

$$2C_{10}H_{13}N + 3[O] \rightarrow C_{20}H_{24}N_2O_2 + H_2O$$

He obtained, not the quinine he sought, but a dirty brown precipitate. Undismayed, he decided to investigate the oxidation of the simpler amine, aniline. After preparing the sulphate of aniline he oxidized it with potassium dichromate and obtained a black precipitate which, after drying and extraction with alcohol, yielded a brilliant purple solution. This product proved to be an extremely good dyestuff and it was rapidly accepted by British and French dyers. It is interesting to note that the dye was as costly as platinum and that by the time theoretical knowledge had progressed far enough to elucidate its structure in 1888, the dye had fallen out of general use, finding its last major application in 1881 for the

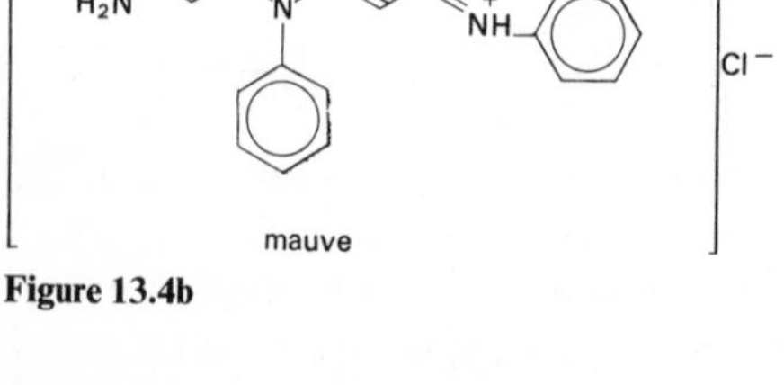

mauve

Figure 13.4b

printing of lilac coloured 1d postage stamps. (Figure 13.4b shows the structure of the dye known as Perkin's mauve; a photograph of the stamp is shown inside the back cover of *Students' Book I*.)

Hofmann forecast that Britain would become the chief dye manufacturing and exporting country in the world because of the ready availability of the starting material for dyestuffs, coal tar. As we shall see, this did not prove to be an accurate forecast. In the first twenty years of the dye industry the inventive genius in synthesizing new aniline based dyes lay almost wholly in Britain and France. German chemists sought experience in Britain, and many involved in this 'brain drain' made enormous contributions to the field. However, the vast British commitment in the textile, coal, and iron industries of the industrial revolution overshadowed the growth of the dyestuffs industry which began to take firm roots in Germany. When the First World War broke out, Britain was importing a large proportion of its dyes from Germany. With the cessation of imports in 1914 British dyers were so deprived that they had insufficient dyestuffs to dye the uniforms of the troops who were to fight the Germans! So acute was the shortage that Royal Warrants were issued to permit trading with the enemy and dyes were purchased from Germany for a while by way of Holland.

There are some modern instances of this sort of thing: the development of fundamental new techniques by the scientists and engineers of one country, and their subsequent exploitation by others. In this category we might include the computer, penicillin, and the hovercraft.

One of the most important contributions made by German scientists to the dyestuff field was the discovery by Griess, in 1858, of the diazotization reaction.

You have seen the reaction between a primary amine, such as butylamine, $C_4H_9NH_2$, or aniline, C_6H_5—NH_2 and an acid, resulting in the formation of a salt. Hydrochloric acid, and aniline, for example, react in the following way.

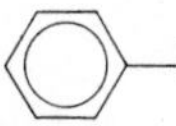

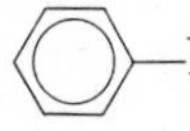

$$C_6H_5\text{—}NH_2 + H^+Cl^-(aq) \rightarrow C_6H_5\text{—}NH_3^+Cl^-(aq)$$

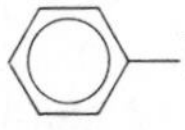

The reaction with one acid, nitrous acid, HNO_2, is unusual and complicated. Again taking aniline as an example, if the reaction takes place at room temperature, some phenol is formed and nitrogen is evolved.

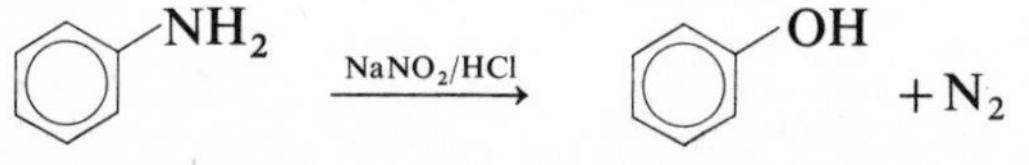

$$C_6H_5NH_2 \xrightarrow{NaNO_2/HCl} C_6H_5OH + N_2$$

If the temperature is kept low by cooling in an ice bath, no nitrogen is evolved and benzenediazonium chloride is formed in solution.

$$C_6H_5NH_2 \xrightarrow[5\,^{\circ}C]{NaNO_2/HCl} C_6H_5N_2^+Cl^-$$

This is the diazotization reaction. It takes place with all primary aromatic amines such as aniline. Benzenediazonium chloride, if isolated, crystallizes as an explosive salt, but it is safe in solution where it can be used as a valuable intermediate for further synthesis.

One important reaction of benzenediazonium chloride is its ability to couple with phenols and with aromatic amines to give brightly coloured azo-compounds. Two examples are:

1 Coupling with phenol:

$$C_6H_5{-}N_2^+Cl^- + C_6H_5{-}OH \xrightarrow{NaOH} C_6H_5{-}N{=}N{-}C_6H_4{-}OH$$

p-hydroxyazobenzene (orange)

2 Coupling with dimethylaniline:

$$C_6H_5{-}N_2^+Cl^- + C_6H_5{-}N(CH_3)_2 \xrightarrow{H_2O} C_6H_5{-}N{=}N{-}C_6H_4{-}N(CH_3)_2$$

p-dimethylaminoazobenzene (yellow)

The colour is due to the presence of the —N=N— group in conjunction with the benzene rings. A group which confers colour is known as a chromophore.

Not all dyestuffs are azo-dyes, of course, but all do contain chromophoric groups of some kind. Here are some examples of azo-dyes in current use; the preparation of the sodium salt of methyl orange or of 'Dispersol' Fast Yellow is one that you could undertake yourself.

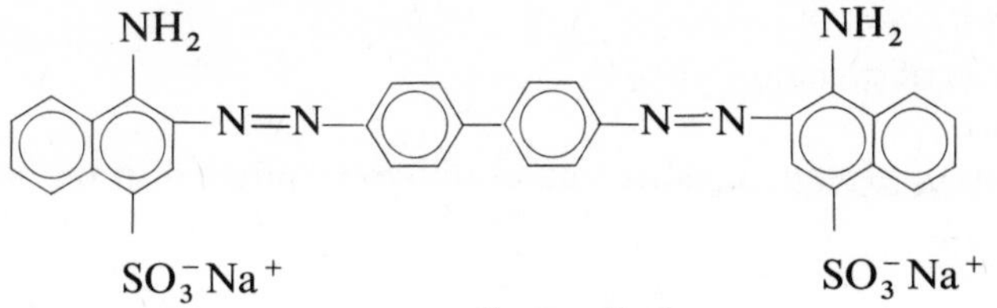

NH_2 NH_2

—N=N— —N=N—

$SO_3^-Na^+$ $SO_3^-Na^+$

Congo Red

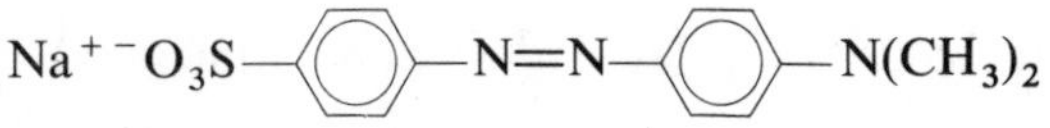

$Na^{+}\,{}^{-}O_3S$— —N=N— —$N(CH_3)_2$

Sodium salt of methyl orange

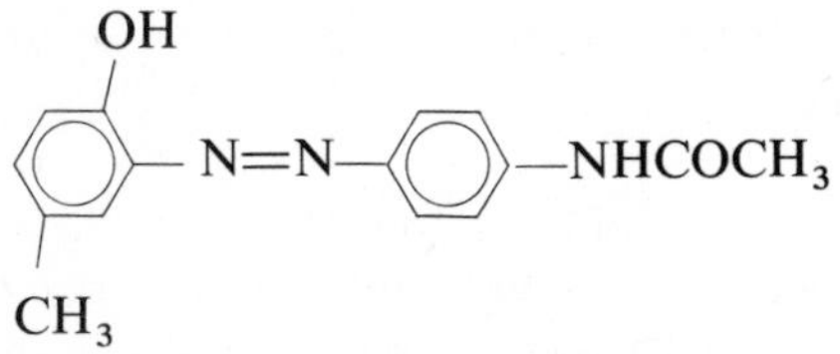

OH

—N=N— —$NHCOCH_3$

CH_3

'Dispersol' Fast Yellow G

The point to be borne in mind here is that there is a link between colour and constitution, between the structure of a molecule and the properties we desire. This seeking for, and exploitation of, the relationships between structure and properties is fundamental to the operation of the organic chemical industry.

13.5 Two different functional groups in the same molecule

In this section you are going to investigate the effect of having two different functional groups in the same molecule. A suitable compound to investigate is glycine, or aminoacetic acid. Its structural formula is

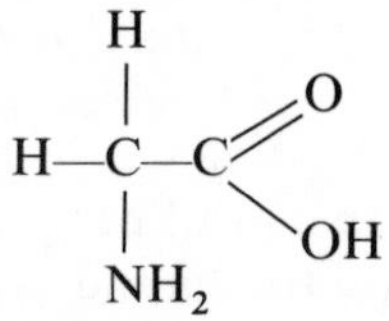

H

H—C—C(=O)—OH

NH_2

and it includes two functional groups that you have already studied, the —NH_2 group of amines, and the —CO_2H group of carboxylic acids.

Write down some of the properties that you think this compound might have, and then do the experiment.

Experiment 13.5a

Properties of glycine

Procedure

Make a solution of glycine in water and examine the effect of it on universal indicator solution.

In the presence of the indicator, add to separate portions of the solution, *in drops* from a teat pipette, some very dilute acids and alkalis and observe the effect.

Add, in drops, some glycine solution to a few cm^3 of copper(II) sulphate solution.

What do these reactions tell us about the properties of glycine?
How do you think glycine might behave on electrolysis (*a*) in acid solution, (*b*) in neutral solution, and (*c*) in alkaline solution? Test your conclusion as follows.

Cut three pieces of filter paper to the size of a microscope slide. Place one of these on a slide and secure it in place at each end by means of a crocodile clip to which is attached a length of wire to connect to a source of d.c. of about 50 volts.
Note. It is most important that you hold the filter paper only *at the edges.* Holding it elsewhere will obscure the result, as is explained later.

Do the same thing with the other two pieces of filter paper. Moisten one of them with dilute acetic acid, one with water, and one with ammonia solution. By means of a thin wire, place a very thin streak of glycine solution across the middle of each filter paper, parallel with the short side of the slide. Mark the position in pencil (not ballpoint or ink),mark + and − on the appropriate ends of the filter paper, and mark the paper (*a*), (*b*), or (*c*) as above. Connect each slide to the 50 volt d.c. supply using the crocodile clips and set aside for 20 minutes.

At the end of this time disconnect the apparatus and holding the filter paper with tongs spray it with a ninhydrin aerosol spray, or dip it in a 1 per cent solution of ninhydrin in acetone. Allow the papers to dry in the air and then heat them in an oven at about 100 °C for about 10 minutes. (If you have no oven available, hold them with tongs in front of an electric fire, or a few feet above a Bunsen burner, but be careful not to scorch the paper.)

After this time the position of the glycine streaks will be made visible as a purple coloration. The glycine will be found to have moved from its original position. This movement is known as electrophoresis.

All amino acids on heating with ninhydrin give coloured compounds, most of which are purple. If you have held the paper other than at the edges, you will have given it a 'finger print' which contains amino acids, and a purple patch will appear on heating, confusing the result.

Background reading

Medical applications of electrophoresis

Electrophoresis is proving very valuable in clinical chemistry for diagnosing certain diseases.

The method is employed for rapid separation of compounds which would otherwise be difficult to separate, or which could not be separated at all. It is especially suited to separation of the various protein fractions which are present in blood. In a number of diseases changes occur in these protein fractions, and recognition that the changes have occurred is of considerable importance in diagnosing the diseases.

As a supporting medium, paper has been largely replaced by cellulose acetate strip. This is saturated with a solution of constant acidity (a buffer solution) before being placed between the electrodes. Separation of the protein fractions of blood serum on this medium is rapid, usually about 75 minutes, and staining and washing takes a further 15 minutes. This pattern obtained is greatly superior to that given by a paper medium.

Normal blood serum shows a dense band of albumen which has migrated most rapidly towards the anode, and this is followed by less intense bands of α- and β-globulin fractions; γ-globulins migrate towards the cathode. Albumen is the main protein of blood, and is manufactured by the linking together of amino acids in the liver. The globulins are a diverse group of proteins with a wide range of functions. α-globulins carry iron; β-globulins carry fatty substances; and the γ-globulins, which are of very high molecular weight, include all the antibodies. Antibodies coagulate and precipitate toxins in the blood, and are part of the body's defences against infection.

In some diseases, for instance those of the liver or kidney, there are marked increases in the α- or in the γ-globulin fractions. In others, the γ-globulin fraction is decreased or even absent. The appearance of a dense band of abnormal protein is strong evidence for a particular disease of bone marrow.

Electrophoresis can also be used to separate normal haemoglobin from the abnormal haemoglobins which occur in certain anaemic conditions, and identification of the abnormal constituents can lead to diagnosis.

Electrophoresis therefore provides a rapid and simple means of separating complex and often delicate molecules that could not readily be separated by other means, and it is proving of great value in clinical medicine.

Asymmetry of structure

Many amino acids have asymmetric molecules. The meaning of this term will be explained in class, when the properties of amino acids, seen in experiment 13.5a, are discussed. In the next experiment you will investigate an interesting effect caused by the asymmetry of molecular structure of such compounds.

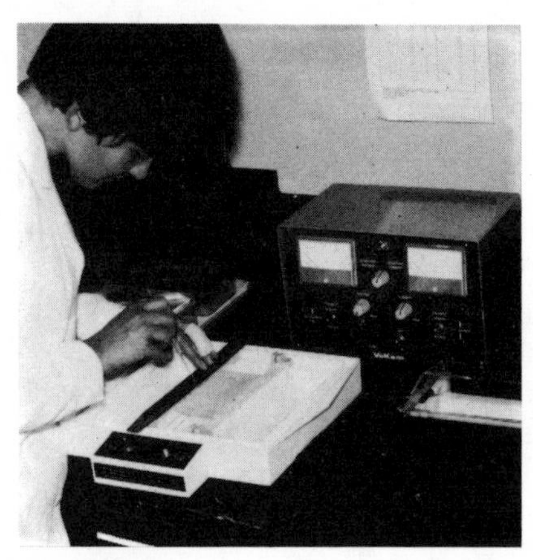

Figure 13.5a
Apparatus for electrophoresis.
Shandon Scientific Company Ltd

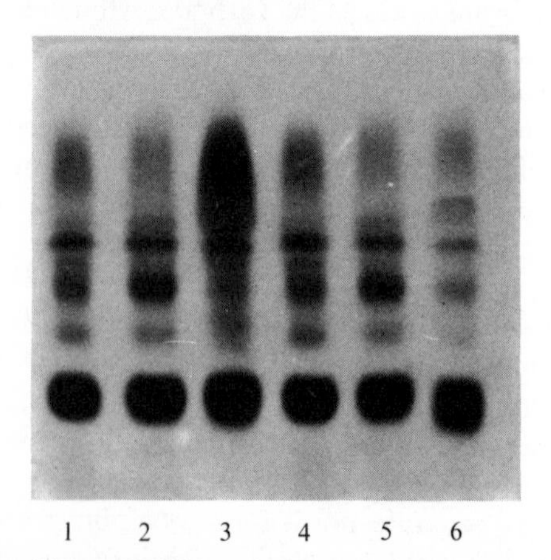

1 2 3 4 5 6

Figure 13.5b
Serum proteins separated by electrophoresis on cellulose acetate: 1 Normal, 2 Acute inflammation, 3 Cirrhosis of the liver, 4 Normal, 5 Diabetes, 6 Bisalbuminaemia.

Experiment 13.5b

To find out if polarized light is affected by asymmetric isomers of compounds

In Topic 8 you investigated the effect certain crystals had on polarized light. You are now going to find out if *molecules*, as opposed to a *crystal structure*, can show a similar effect. To obtain separate molecules in a non-crystalline arrangement, we must have a gas, a liquid, or a solution.

You will be given two solid amino acids, the names and structural formulae of which are:

L-cystine

$$\begin{array}{c} \quad\quad\ \ \mathrm{H} \quad\quad\quad\quad\quad\quad\quad\quad\quad\quad\quad\ \ \mathrm{H} \\ \quad\quad\ \ | \quad\quad\quad\quad\quad\quad\quad\quad\quad\quad\quad\ \ | \\ \mathrm{HO_2C{-}C{-}CH_2{-}S{-}S{-}CH_2{-}C{-}CO_2H} \\ \quad\quad\ \ | \quad\quad\quad\quad\quad\quad\quad\quad\quad\quad\quad\ \ | \\ \quad\quad\ \ \mathrm{NH_2} \quad\quad\quad\quad\quad\quad\quad\quad\quad\ \ \mathrm{NH_2} \end{array}$$

and

L-glutamic acid

$$\begin{array}{c} \quad\quad\quad\quad\quad\quad\quad\quad\quad\ \mathrm{H} \\ \quad\quad\quad\quad\quad\quad\quad\quad\quad\ | \\ \mathrm{HO_2C{-}CH_2{-}CH_2{-}C{-}CO_2H} \\ \quad\quad\quad\quad\quad\quad\quad\quad\quad\ | \\ \quad\quad\quad\quad\quad\quad\quad\quad\quad\ \mathrm{NH_2} \end{array}$$

You will also have access to a polarimeter. The diagram on the next page shows how this instrument is used.

Make a solution of 0.5 g of L-cystine in 10 cm^3 of approximately M hydrochloric acid. Place half of it in a specimen tube of about 12 mm diameter and take it to the polarimeter.

Adjust the polarimeter by rotating the centre of the analyser until, on looking through the analyser and polarizer, the source of light is extinguished. Note the position of the pointer on the scale. Place the specimen tube in position and look through the instrument once more. Do you have to alter the setting of the analyser to extinguish the light, and if so, by how much?

Now add the other half of the solution to that already in the specimen tube so as to double the length of liquid through which the light passes. Is a further adjustment of the analyser necessary for extinction?

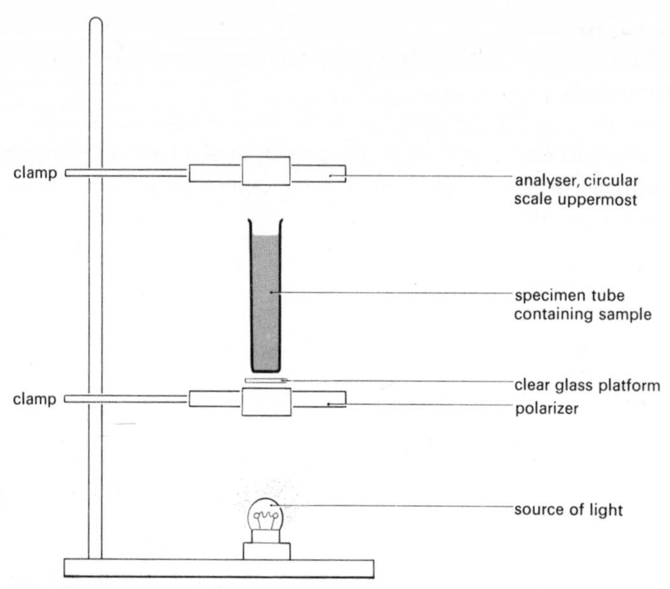

Figure 13.5c
A simple polarimeter.

Now dissolve another 0.5 g of cystine in the mixture in the specimen tube, so that you double the concentration of the solution. What effect does this have?

Can you work out a tentative relation between the angles of rotation and (*a*) the length of column of solution, and (*b*) the concentration of the solution?

Now repeat the whole experiment with L-glutamic acid. You will be able to dissolve up to about 2 g of the acid in 10 cm^3 of M hydrochloric acid. Begin by drawing up a plan of how you will proceed. You should aim at obtaining several values of angles for different lengths of column of liquid and concentration of solution. Do your results confirm the relationship you have previously suggested?

13.6 **A problem in synthesis**

To conclude this part of your study of carbon compounds it is suggested that you should attempt to plan and execute the synthesis of a compound of your own choice. A week is a reasonable length of time for you to spend on this work so the synthesis should not include more than two or three separate reactions, e.g.

$$\text{an alcohol} \xrightarrow{\text{dehydration}} \text{an alkene} \xrightarrow{\text{halogenation}} \text{a dihalogenoalkane}$$

Practical books will give you some ideas about synthetic routes which it might be possible to follow and then a compound can be chosen which could be synthesized by one of the routes you have noted.

You may have read of a compound earlier in this topic or in Topic 9 which you would like to synthesize; as an alternative you might select a compound with a disubstituted benzene ring, such as NO_2 (benzene ring) $-NH_2$ and attempt its synthesis from benzene.

The planning of synthetic routes is an important activity of organic chemists and you should check all details with care, including a consideration of each of the following items.

1 The scale of operations. With a 60 per cent yield at each stage, how much product can you expect from 10 g of starting material after three stages? Assume that there is no change in molecular weight.

2 The cost of raw materials. Check these from a catalogue; you are unlikely to be able to use a starting material costing £3 for 100 g if an alternative route has a starting material costing only 25 p.

3 Can the reactions be carried out with the available apparatus and in the limits of your technical ability?

4 Are the chemicals you propose to handle harmless? If they are toxic, would a fume cupboard give adequate protection or would it be wiser to drop the project?

In spite of all the care you take in planning there may be difficulties that you are unaware of through inexperience and you must consult your teacher before starting your synthesis.

Problems

1 A substance, A, had the molecular formula C_3H_7Br. After boiling with aqueous sodium hydroxide a compound, B, of molecular formula C_3H_8O, was formed. On oxidation B formed C (C_3H_6O) which gave a precipitate with 2,4-dinitrophenylhydrazine, but had no reaction with Fehling's solution. Name and give the structural formulae of A, B, and C.

2 A substance, A, had the molecular formula $C_4H_{10}O$. On oxidation it gave B (C_4H_8O) which gave a precipitate of copper(I) oxide with Fehling's solution.

On passing the vapour of A over heated silica it formed C (C_4H_8). C reacted with hydrogen iodide to form D (C_4H_9I). D after hydrolysis and then oxidation formed E which gave a precipitate with 2,4-dinitrophenylhydrazine, but had no reaction with Fehling's solution. Name and give the structural formulae of A, B, C, D, and E.
Give the name and structural formula of one isomer of A which would also undergo this series of reactions.

3 A substance, A, (molecular formula $C_5H_{13}N$) existed as two enantiomorphs. On treatment with hydrochloric acid, A formed B ($C_5H_{14}NCl$). A also reacts with acetyl chloride to form C ($C_7H_{15}NO$).

i Give structural formulae which could apply to A, B, and C.

ii Give the structural formulae of two isomers of A which do not exist as enantiomorphs.

4 Aspartic acid has the formula

$$\begin{array}{c} HO_2C\text{—}CH_2\text{—}\underset{\displaystyle NH_2}{\underset{|}{CH}}\text{—}CO_2H \end{array}$$

i State how the formula
 a resembles
 b differs from that of glycine.

ii Deduce how the properties of aspartic acid
 a resemble
 b differ from those of glycine.

5 Ether is a valuable solvent for certain types of carbon compound but is very volatile, a factor which restricts its use. Another similar compound, tetrahydrofuran,

$$\begin{array}{ccc} H_2C & \text{—} & CH_2 \\ | & & | \\ H_2C & & CH_2 \\ & \diagdown\, O \,\diagup & \end{array}$$

, is an equally effective solvent, and is not so volatile.

A certain laboratory required a similar solvent but one with an even higher boiling point, and decided to attempt to make a derivative of tetrahydrofuran having the following structure:

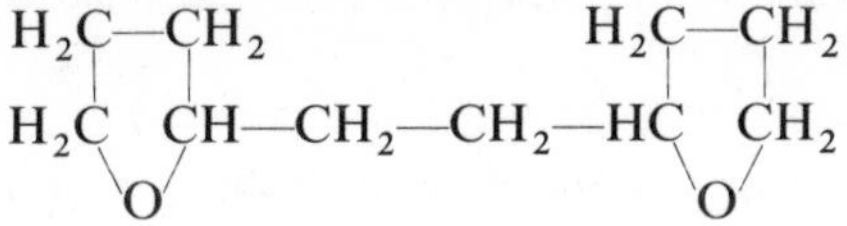

```
H2C—CH2                 H2C—CH2
 |    |                  |    |
H2C  CH—CH2—CH2—HC   CH2
  \  /                   \  /
   O                      O
```

There was available a supply of furfuryl alcohol,

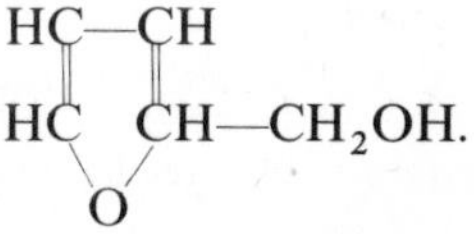

```
HC—CH
‖    ‖
HC   CH—CH2OH.
  \  /
   O
```

A survey of textbooks revealed the following general reactions of functional groups that were considered to be relevant:

i $>C=C<$ would undergo addition reactions

a with hydrogen catalytically to make $>CH—CH<$

b with halogen to make $>CX—CX<$

c with hydrogen halide to make $>CH—CX<$

d with sulphuric acid to make $>CH—CHSO_4$

e with water, catalytically, to make $>CH—COH$

ii $—CH_2OH$ can be converted
a to $—CHO$ and to $—CO_2H$ by oxidation
b to $—CH_2X$ by several methods, e.g. to $—CH_2Cl$ using $SOCl_2$

iii $—CH_2X$ can be converted to
a $—OH$ by alkali
b $—OR$ by $Na^{+}\,{}^{-}OR$
c $—CN$ by $K^{+}CN^{-}$
d RX can be changed to R—R by the action of sodium, the equation for which is $2RX+2Na \rightarrow R—R+2NaX$

You are asked to devise a route by which this laboratory might attempt to convert furfuryl alcohol into the desired end product. You may use the reactions mentioned above, or any others known to you. Suggest the best possible route to try first, and write the full structural equations for all the reactions you mention.

6 When a halogenoalkane, such as 1-bromopropane, is heated with potassium hydrogen sulphide in ethanol solution, a compound known as propane-1-thiol is formed

$$CH_3CH_2CH_2Br + KSH \rightarrow CH_3CH_2CH_2SH + KBr$$

As oxygen and sulphur are in the same group of the Periodic Table, it is reasonable to suppose that the —SH and —OH groups of atoms might have similar properties. Assuming this to be the case, from your knowledge of the chemistry of alcohols, suggest possible methods of making the compounds having the structures listed below. The only carbon compound you may start with is 1-bromopropane.

$CH_3CH_2CH_2S^-Na^+$
$CH_3CH_2CH_2SCH_2CH_2CH_3$
$CH_3CH_2COSCH_2CH_2CH_3$

7 The remaining questions also involve the conversion of one compound into another, but do not give any guidance on how this should be done. If you cannot think of a suitable method, you should look in a textbook for likely reactions that would achieve the result.

How would you carry out the following changes?

i Ethyl acetate from ethanol using no other carbon compound.

ii Two isomeric esters of formula $CH_3CO_2C_4H_9$ starting from ethanol, butan-1-ol, and butan-2-ol.

iii *o*-Cresylbenzoate from *o*-cresol and benzoic acid.

iv Propylamine ($CH_3CH_2CH_2NH_2$) from propan-1-ol.

Topic 14

Reaction rates

14.1 Introduction

If you were a rocket engineer looking for a suitable solid fuel, you might search through the thermochemical literature for a solid that, when burnt, produced a large amount of gas and was very exothermic. Looking through a list of heats of combustion, you might stop at sucrose:

$$\underset{\text{sucrose}}{C_{12}H_{22}O_{11}(s)} + 12O_2(g) \rightarrow 12CO_2(g) + 11H_2O(l); \Delta H^{\ominus}_{\text{combustion, 298}} = -5644\,\text{kJ mol}^{-1}$$

From this information you might expect sucrose to burst into flames or even to explode in contact with atmospheric oxygen. In fact this substance, which was in the sugar bowl on your breakfast table this morning, is very stable at normal temperatures. Therefore, to decide about its suitability as a rocket fuel, you will want to know at what speed it reacts at various temperatures, as although the rate at room temperature is negligible, at other temperatures it may be different.

Obviously the rate of a reaction will interest us, but once we have found out that it proceeds at a certain rate, what else can we do? We shall see that a study of the factors upon which the rate depends can tell us something of what occurs between the reactants and the products. If warm, acidified solutions of permanganate and oxalate ions are mixed, a reaction occurs:

$$2MnO_4^-(aq) + 16H^+(aq) + 5C_2O_4^{2-}(aq) \rightarrow 2Mn^{2+}(aq) + 10CO_2(g) + 8H_2O(l)$$

At the start of the reaction there are permanganate, hydrogen, and oxalate ions, and water only, and, if the right proportions are chosen, manganese(II) ions, carbon dioxide, and water only are left at the end of the reaction. But what has happened in between? Surely the 23 ions that are represented on the left of the equation did not collide all at once and out of this collision came the 20 particles represented on the right of the equation? A study of the rate of the reaction involves a study of how the reaction occurs: in this case it would be reasonable to assume that the reaction is made up of a set of simpler processes. A model of how the reaction occurs is called its mechanism, and if something can be found out about it, not only has some more of the chemistry of these substances been learned, but it may be possible to determine how the rate can be altered. It is not much good using an industrial process which is extremely efficient and turns 99 per cent of the reactants into products with no side reactions, if it proceeds at a very slow rate.

It will be useful to investigate how various factors affect the rate of a reaction, but before this can be done, what is meant by 'the rate of reaction' must be decided, and also how it can be measured. The rate of the reaction is a measure of how the extent of the reaction changes with time, so in order to measure it, some property which alters with the extent of the reaction must be measurable. In the reaction above, it can be seen that the concentrations of permanganate, hydrogen, and oxalate ions decrease in proportion to the extent of the reaction. Also the concentrations of manganese(II) ions, carbon dioxide, and water increase during the reaction. It would be difficult to measure the concentrations of the latter three, but simple titrimetric procedures for determining the concentrations of permanganate or hydrogen ions are known. The titration would have to be carried out very quickly indeed to find out the concentration at a particular time, if the reaction were proceeding with any speed; once all the hydrogen or permanganate ions present had been titrated, we would have 'killed' the experiment as there would be no more reaction!

This leads to two possibilities. Either several separate experiments can be conducted, each one being stopped after a different interval of time, to find the comparative extents of the reaction. Or, one experiment can be done, samples being removed at intervals. These samples, rather than the whole of the reaction mixture, can be titrated. But then there must be some way of slowing down the reaction in the sample to give time for it to be titrated.

What else changes with the extent of the reaction?

1 The colour of the solution changes during the reaction. The only coloured substance present is the permanganate ion, and its concentration and hence intensity of colour is decreasing during the reaction. The intensity could be followed by a photoelectric device in a colorimeter as shown in figure 14.1a.

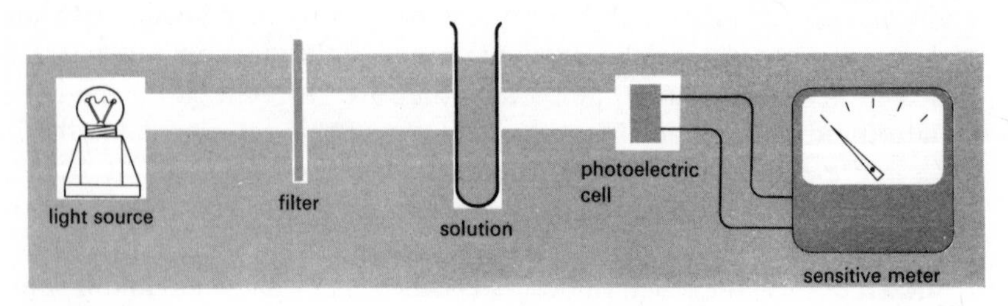

Figure 14.1a
A colorimeter.

2 The total volume of the solution may change. It is difficult to say by looking at the stoichiometric equation whether this will occur; if it does, then by enclosing the reacting system in a bulb with a stem having a small bore, we should be able to detect small changes in the volume of the system. This type of apparatus is known as a dilatometer, and is illustrated in figure 14.1b.

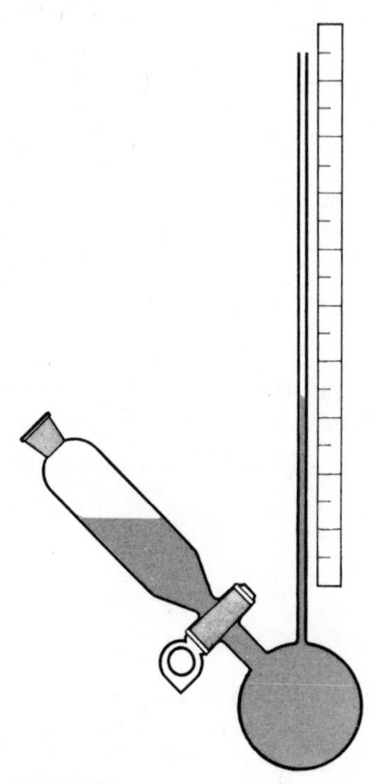

Figure 14.1b
A dilatometer.

3 Carbon dioxide gas is produced during the reaction. If the conditions can be so arranged that it is all evolved, it could be collected, and its volume measured, in an apparatus such as that shown in figure 14.1c.

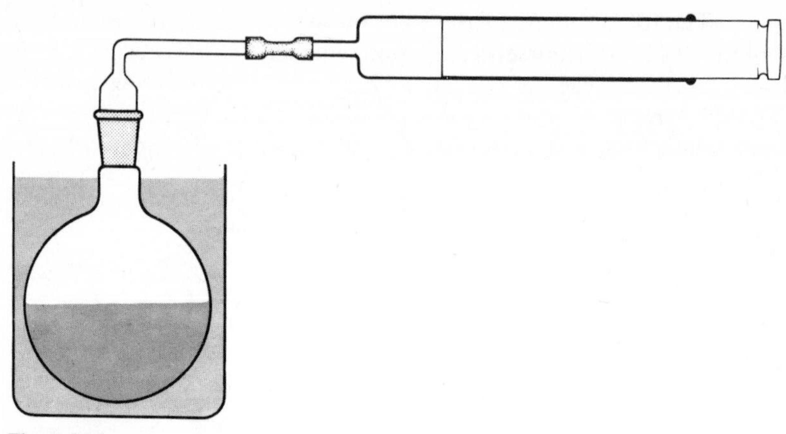

Figure 14.1c
Measurement of gas evolved.

4 As there are 23 ions at the beginning of the reaction and only two at the end, the electrical resistance of the solution will change during the reaction. This can be measured by immersing two electrodes in the solution using an apparatus such as that shown in figure 14.1d.

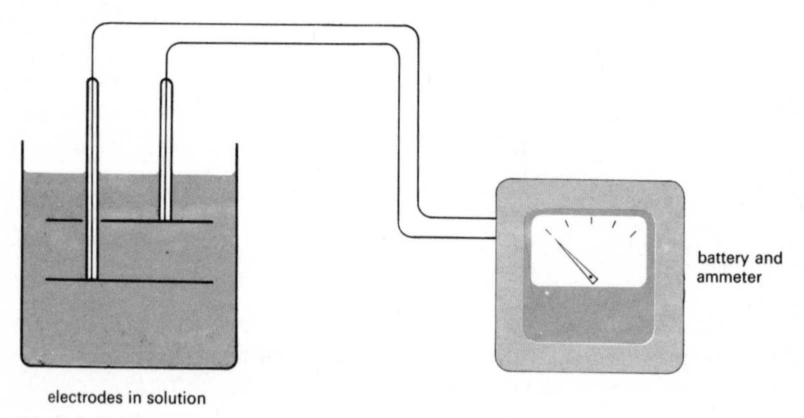

Figure 14.1d
Measurement of the conductivity of a solution.

One great advantage of all the last four methods is that samples need not be taken, thus the extent of the reaction can be determined at intervals of time by some external method without disturbing the system. Not all of these methods may be applicable in this case because of one difficulty or another (for instance, the carbon dioxide dissolves in the solution rather than being evolved). In this case, probably the most convenient method is the colorimetric one.

To sum up, the methods of following the extent of the reaction are:
Sampling, and quenching if necessary
  a Titrimetric
Methods involving external observation
  a Colorimetric
  b Dilatometric (following volume change in a dilatometer)
  c Collecting gas evolved
  d Measuring the electrical resistance of the solution
and others such as
  e Measuring the optical rotation of the solution
  f Measuring the refractive index of the solution

We can describe the rate of the reaction as the rate at which the extent of the reaction proceeds with time. A simple way of expressing this is as the rate of increase of concentration of one of the products of the reaction or the rate of decrease of concentration of one of the reactants. If this is taken as a definition, whenever a rate is quoted it must also be stated *to what concentration the rate refers.* For in the reaction above:

	rate of decrease of concentration of permanganate ions	÷ 2
=	rate of decrease of concentration of hydrogen ions	÷ 16
=	rate of decrease of concentration of oxalate ions	÷ 5
=	rate of *increase* of concentration of manganese(II) ions	÷ 2
=	rate of *increase* of concentration of carbon dioxide	÷ 10
=	rate of *increase* of concentration of water	÷ 8

The rate referred to the increase of concentration of carbon dioxide is therefore five times that referred to the decrease of concentration of permanganate ions.

In spite of these complications, we can use the rate of increase of concentration of products or decrease of concentration of reactants as a useful definition of the rate of a reaction, provided that we state to what concentration it refers. The units of the rate of change of concentration with time are mol dm^{-3} s^{-1}.

14.2 The kinetics of the reaction between iodine and acetone in aqueous solution

In this section we shall investigate the factors which influence the rate of one particular reaction. The conclusions that are reached in this experiment can then be tested by applying them to other reactions.

Experiment 14.2

An investigation of the rate of the reaction between iodine and acetone in aqueous solution

You are going to investigate the reaction between iodine and acetone, the equation for which is:

$$CH_3COCH_3(aq) + I_2(aq) \rightarrow CH_3COCH_2I(aq) + H^+(aq) + I^-(aq)$$

Procedure

Put a few cm^3 of acetone in a test-tube and add a few drops of 0.05M iodine solution. Observe what happens; then add about 5 cm^3 bench dilute hydrochloric acid and observe what happens.

You will know from earlier work that changes of concentration may affect the rate of reactions. We shall now investigate how the concentrations of acetone, of iodine, and of hydrogen ions affect the rate of this reaction. In order to do this we will determine the initial rate of the reaction starting with different concentrations of acetone, iodine, and hydrogen ions. A convenient method of determining the rate of this reaction is to measure the rate of disappearance of iodine using a colorimeter. Use one of the mixtures set out in table 14.2a in such a way that the class covers the full range and the results can be compared. The volumes have been calculated so that in (a), (b), and (c) the concentrations of iodine and of hydrogen ions remain constant and that of acetone changes; in (a), (d), and (e) the concentrations of acetone and of hydrogen ions remain constant and that of iodine changes; and in (a), (f), and (g) the concentrations of acetone and iodine remain constant and that of hydrogen ions changes.

1 Put the iodine solution, acid, and water into a test-tube that fits the colorimeter, using a burette to deliver the correct quantity of each liquid.

2 Put the acetone solution into another test-tube, again measuring from a burette.

3 Add the acetone solution to the other solution, *stopper the test-tube,* mix the contents by rapid shaking, and start the clock.

4 Adjust the meter of the colorimeter to maximum with a tube of water in place at about 15 seconds before each minute.

5 Put the test-tube of the reaction mixture into the colorimeter and note the meter reading at each minute. Continue taking readings for up to 6 minutes.

6 Finally, find the meter reading at time zero by making up a mixture containing water instead of acetone solution and placing this in the colorimeter. Note the meter reading which is obtained.

	Volume of 2M acetone /cm^3	Volume of 0.01M iodine /cm^3	Volume of 2M HCl /cm^3	Volume of water /cm^3	Total volume /cm^3	[Acetone] /mol dm^{-3}	[I_2] /mol dm^{-3}	[H^+] /mol dm^{-3}
a	2	2	2	4	10	**0.4**	**0.002**	**0.4**
b	4	2	2	2	10	**0.8**	0.002	0.4
c	6	2	2	0	10	**1.2**	0.002	0.4
d	2	4	2	2	10	0.4	**0.004**	0.4
e	2	1	2	5	10	0.4	**0.001**	0.4
f	2	2	4	2	10	0.4	0.002	**0.8**
g	2	2	6	0	10	0.4	0.002	**1.2**

Table 14.2a

From the colorimeter readings determine the molarity of the solution with respect to iodine at various times, and then plot a graph with these molarities on the vertical axis against time (in seconds) on the horizontal axis. The *rate of decrease of concentration* of iodine (which is positive) is equal to the *rate of change of concentration* of iodine (which is negative) multiplied by -1, and therefore the initial rate of the reaction can be found from the initial slope of the graph. To measure the gradient, draw a tangent to the curve at time = 0, and determine the gradient of this tangent. Many of the graphs will approximate to straight lines, and it may not therefore be necessary to draw tangents to obtain the initial rate. Compare the various rates for the different initial concentrations and so find out if and how the concentrations of the three substances affect the rate of the reaction. In order to compare the rates, we must have some idea of the errors involved. If two rates are

a 1.3 ± 0.3 mol dm^{-3} s^{-1}, and
b 0.8 ± 0.2 mol dm^{-3} s^{-1}

then although inspection of 1.3 and 0.8 mol dm^{-3} s^{-1} might suggest that the reaction in (a) was nearly twice as fast as that in (b), the errors indicate that (a) can be as low as 1.0 mol dm^{-3} s^{-1} and (b) as high as 1.0 mol dm^{-3} s^{-1}, so the rate in the two experiments may be the same. Comparison with other students' results will enable you to gain some idea of the errors, and, indeed, an average of several experimental results will give better figures with which to decide whether the rate is being affected.

Using the experimental results of the whole class, try to answer these questions:

1 Is the rate of the reaction affected by the concentration of acetone?
2 How is it affected?
3 Can you express this mathematically?

Plot a graph of rate against concentration of acetone.

4 Can you obtain a value for the constant in the mathematical expression in question 3 from this graph?

5 What are the units of this constant?

6 Is the rate of the reaction affected by the concentration of iodine?

7 Is it affected by the hydrogen ion concentration?

8 Can you express this mathematically?

9 Bearing questions 3 and 8 in mind, can you suggest an equation that describes how the rate of this reaction depends on the concentrations of reactants?

10 What are the units of the constant in your expression?

You will be discussing your answers to these questions in class, and arriving at a mathematical expression connecting the rate of the reaction at any instant with the concentration of the reactants present at that time.

Some of the terms in this expression have special names, and for the purposes of defining these terms let us now consider the general case of a reaction having the equation

$$mA + nB \rightarrow \text{products}.$$

Let us suppose that the rate is found by experiment to be proportional to $[A]^x$ and $[B]^y$.

We can therefore write

$$\text{Rate} = k[A]^x[B]^y$$

and this equation is known as the *rate expression* for the reaction. The constant k is known as the *rate constant*, and is a constant for a given reaction at a particular temperature. It is found by experiment however to vary with temperature.

The indices x and y are known as *orders*. The reaction is said to be of x order with respect to A, y order with respect to B, and $(x+y)$ order overall.

The order of a reaction is thus the sum of the indices of the concentration terms in the experimentally-determined rate expression.

Now you have investigated the factors which influence the rate of one reaction, you can apply your knowledge to some other reactions in the following problems.

Problem 1

Table 14.2b contains some data for the decomposition of dinitrogen pentoxide in tetrachloromethane solution.

$$2N_2O_5(\text{in } CCl_4 \text{ solution}) \rightarrow 4NO_2(\text{in } CCl_4 \text{ solution}) + O_2(g)$$

Concentration N_2O_5 /mol dm^{-3}	**Rate of reaction as decrease in concentration of N_2O_5 per second /10^{-5} mol dm^{-3} s^{-1}**
2.21	2.26
2.00	2.10
1.79	1.93
1.51	1.57
1.23	1.20
0.92	0.95

Table 14.2b

Plot a graph of the rate of the reaction against the concentration of N_2O_5, and try to answer these questions:

1 What is the rate expression for the reaction?
2 What order is this reaction with respect to N_2O_5?
3 What is the value of the constant in the rate expression? State units.

Problem 2

Table 14.2c contains some results for the reaction:

$$2H^+(aq) + H_2O_2(aq) + 2I^-(aq) \rightarrow I_2(aq) + 2H_2O(l)$$

y is a measure of the concentration of hydrogen peroxide at time *t* seconds. (The iodide ion concentration was kept constant in this experiment.)

y	$t/10^3$ s
20.95	0
18.95	0.562
16.95	1.192
14.95	1.901
12.95	2.714
10.95	3.667
8.95	4.805
6.95	6.233
4.95	8.151
2.95	11.072
0.95	17.471

Table 14.2c

(These results are taken from a famous paper published in 1867 by A. V. Harcourt and W. Esson, after whom this reaction is named.)

Plot a graph of y (on the vertical axis) against t (on the horizontal axis) and from it obtain gradients at four concentrations. Plot these rates against concentration and so verify that the reaction is first order.

Also determine the time for the concentration of hydrogen peroxide to drop to one half of its initial value, from one half to one quarter, from one quarter to one eighth, and from one eighth to one sixteenth of its initial value.

Problem 3

The following data (table 14.2d) were obtained relating to the decay of iodine-128.

Time /s	**Counts of activity of sample /s^{-1}**
1020	116.4
1740	85.2
3000	45.6
3600	35.3
4560	21.4
6300	9.7

Table 14.2d

Plot a graph of counts (s^{-1}) against time, and determine the time for the count to drop

a to one half of the first value (that at 1020 s),
b from one half to one quarter, and
c from one quarter to one eighth of the first value.
d What type of kinetics does this radioactive decay follow?

Problem 4

Table 14.2e contains results from an experiment concerning a second order reaction in which 0.0125 moles each of bromoethane and potassium hydroxide in solution are mixed and then 20 cm^3 samples taken and titrated against 0.05M acid. The titres are directly proportional to the concentration of potassium hydroxide.

Time/10^3 s	Titre of acid/cm^3
0	20.0
1.8	10.5
3.6	7.7
7.2	4.7
10.8	3.6
14.4	2.6
18.0	2.2
21.6	1.8

Table 14.2e

Plot a graph of titre against time, and determine the time for the concentration to fall

a to half of its initial value,
b from one half to one quarter, and
c from one quarter to one eighth.

14.3 **The hydrolysis of bromoalkanes**

In Topic 9 you investigated the different rates at which various halogen compounds take part in hydrolysis reactions. In one experiment, for example, the rates of hydrolysis of 1-chloro, 1-bromo, and 1-iodobutane were compared. Examination of the rates of such reactions provides useful evidence on which theories of their mechanisms can be advanced.

In this section we shall investigate the rates of hydrolysis of the three isomeric bromoalkanes of formula C_4H_9Br, in order to find out something about the mechanisms of their reactions.

Experiment 14.3

A comparison of the rates of hydrolysis of some bromoalkanes

The names and formulae of the three compounds to be used are:

1-bromobutane $CH_3CH_2CH_2CH_2Br$
2-bromobutane $CH_3CH_2CHBrCH_3$
2-bromo-2-methylpropane $(CH_3)_3CBr$

You will recall that 1-bromobutane reacts with hydroxide ions, supplied either from water or from, say, sodium hydroxide solution, according to the equation:

$$CH_3CH_2CH_2CH_2Br + OH^- \rightarrow CH_3CH_2CH_2CH_2OH + Br^-$$

To ensure that the halogen compound and water mix together, they are both dissolved in ethanol.

Procedure

In each of the three test-tubes place 1 cm^3 of ethanol. Using separate teat pipettes and working as quickly as possible, in one, place two drops of 1-bromobutane, in the next, two drops of 2-bromobutane, and in the third, two drops of 2-bromo-2-methylpropane. Now place 1 cm^3 of 0.1M silver nitrate solution in each, and quickly shake the tubes to mix the contents. Note and time carefully what you observe throughout the next five minutes, and longer if possible. Record the temperature.

Does the difference in alkyl group have an effect upon the rate of hydrolysis? If so, which is fastest and which slowest?

This hydrolysis might occur in one of two possible ways.

1 The bromoalkane might first ionize

$$\begin{array}{ccc}
CH_3 & & CH_3\\
| & & |\\
CH_2 & & CH_2\\
| & & |\\
CH_2 & \rightarrow & CH_2\\
| & & |\\
C & & C^+ + Br^-\\
H \diagup \ | \ \diagdown Br & & H \diagup \ |\\
H & & H
\end{array}$$

and then the positive ion so formed might react with an OH^- ion to give the alcohol

$$\begin{array}{ccc}
CH_3 & & CH_3\\
| & & |\\
CH_2 & \rightarrow & CH_2\\
| & & |\\
CH_2 & & CH_2\\
| & & |\\
C^+ + OH & & C\\
H \diagup \ | & & H \diagup \ | \ \diagdown OH\\
H & & H
\end{array}$$

2 Alternatively, attack by the hydroxide ion and ejection of a bromide ion might happen in such a way that for an instant of time both the incoming hydroxide and outgoing bromide ions are equally associated with the hydrocarbon group in a *transition state*.

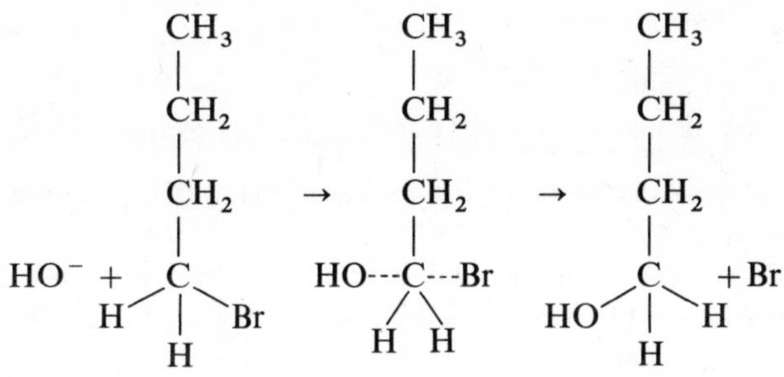

```
          CH3                  CH3                 CH3
          |                    |                   |
          CH2                  CH2                 CH2
          |                    |                   |
          CH2       →          CH2       →         CH2
          |                    |                   |
HO⁻  +    C               HO---C---Br              C      +Br⁻
       H / | \ Br             / \            HO / | \ H
           H                 H   H                 H
```

As the rates of the reactions you have just investigated are very different, it may be that they follow different mechanisms. You will be discussing in class some evidence which makes it possible to decide the point. You may also like to think about possible mechanisms for the reaction between acetone and iodine.

Problem 5

In acid solution, bromate ions slowly oxidize bromide ions to bromine.

$$BrO_3^-(aq) + 5Br^-(aq) + 6H^+(aq) \rightarrow 3Br_2(aq) + 3H_2O(l)$$

Use the following experimental data to determine how the initial rate depends on the concentration of bromate, bromide, and hydrogen ions.

Mixture	**Volume of M bromate /cm^3**	**Volume of M bromide /cm^3**	**Volume of M H^+(aq) /cm^3**	**Volume of water /cm^3**	**Relative rate of formation of bromine**
A	100	500	600	800	1
B	50	250	600	100	4
C	100	250	600	50	8
D	50	125	600	225	2

Table 14.3

Question taken from a scholarship paper set by the University of Cambridge Local Examinations Syndicate.

14.4 Theories of reaction kinetics

1 The collision theory

Many second order reactions in the gas phase

$$A(g) + B(g) \rightarrow \text{products}$$

have rate constants of the same magnitude. Consider the decomposition of a gas A at 500 °C with an initial concentration of A of 0.01 mol dm^{-3} and with a rate constant of 0.01 dm^3 mol^{-1} s^{-1} for the reaction:

$$A(g) + A(g) \rightarrow \text{products}$$

It can be calculated that:

total number of collisions per cubic metre per second $= 3 \times 10^{33}$
number of effective collisions per cubic metre per second $= 3 \times 10^{20}$

which indicates that only one in every 10^{13} collisions results in reaction.

In a chemical reaction, bonds are first broken and then others are made. Energy is therefore required to start this process, whether the reaction is exothermic or endothermic overall. It is reasonable to assume, therefore, that reaction occurs only as a result of those collisions which occur between particles having a certain minimum energy. The fraction of collisions with an energy greater than a value, E, is given by:

$$\lg(\text{fraction}) = -\frac{E}{2.3\,RT}$$

So the fraction with energy greater than 200 kJ mol^{-1} (not quite as large as the energy necessary to break completely a C—C, or C—Cl, or C—O bond) is given by:

$$\lg(\text{fraction}) = -\frac{200 \times 1000}{2.3 \times 8.4 \times 773}$$

from which the fraction $= 2.8 \times 10^{-14}$

This means that one collision in about 4×10^{13} has an energy greater than 200 kJ mol^{-1}. Thus the rate of collisions per cubic metre per second which have an energy greater than 200 kJ mol^{-1}

= total number of collisions per cubic metre per second × fraction of collisions which have an energy greater than 200 kJ mol^{-1}
$= 3 \times 10^{33} \times 2.8 \times 10^{-14}$
$= 8.4 \times 10^{19}$ m^{-3} s^{-1}

which is very similar to the number of effective collisions per cubic metre per second

$= 3 \times 10^{20}$ m^{-3} s^{-1}

It would therefore seem, remembering that the choice of 200 kJ mol^{-1} was arbitrary, that the idea that only those collisions with a certain minimum energy result in reaction is correct. This minimum energy is called the *activation energy*. It can be illustrated as in figure 14.4a.

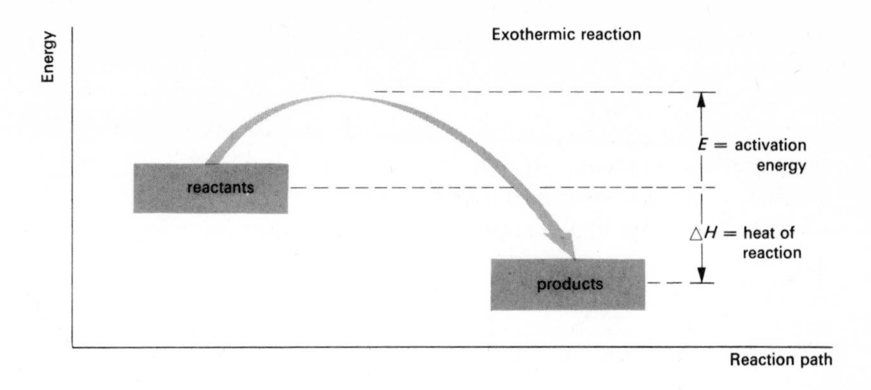

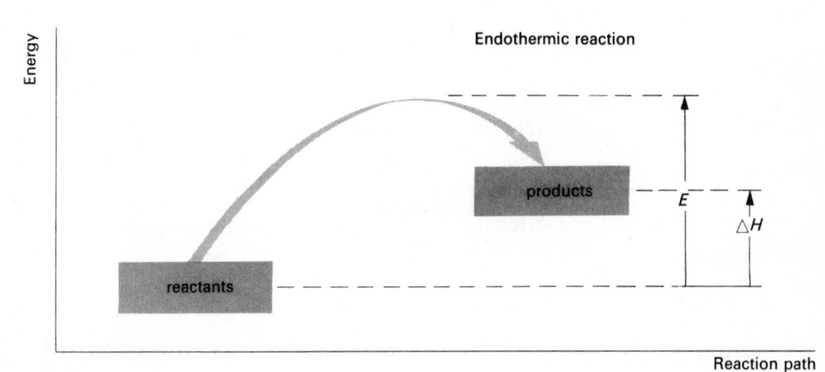

Figure 14.4a
Energy diagrams for exothermic and endothermic reactions.

Experimental confirmation for the theory can be obtained by considering that the rate of effective collision, which is a measure of the rate of reaction, depends on the fraction of collisions having an energy greater than or equal to the activation energy (E).

It can be shown that the rate of reaction depends on temperature according to the relation

$$\text{rate of reaction} \propto 10^{-E/2.3RT}$$

The rate constant k is a measure of the rate of the reaction, independent of concentration, so

$$k \propto 10^{-E/2.3RT}$$

therefore $k = \text{constant} \times 10^{-E/2.3RT}$
and $\lg k = \lg \text{constant} + \lg 10^{-E/2.3RT}$
since the logarithm of a constant is itself a constant

$$\lg k = \text{C} - \frac{E}{2.3R} \times \frac{1}{T}$$

Comparing this with $y = \text{c} + \text{m}x$

shows that a graph of $\lg k$ against $\left(\frac{1}{T}\right)$ will be a straight line, if the theory is valid, and its gradient will be

$$-\frac{E}{2.3\,R}$$

where E = activation energy in joules mol^{-1} and $R = 8.3\ \text{J mol}^{-1}\ \text{K}^{-1}$.

2 The transition state theory

Another method of considering reaction rates is by the *transition state theory*. For the reaction

$$2\text{HI(g)} \rightarrow \text{H}_2\text{(g)} + \text{I}_2\text{(g)}$$

a transition state, $H_2I_2^{\ddagger}$($\ddagger$ is a symbol which indicates a transition state) can be suggested which will be at the maximum of the reaction path as in figure 14.4b.

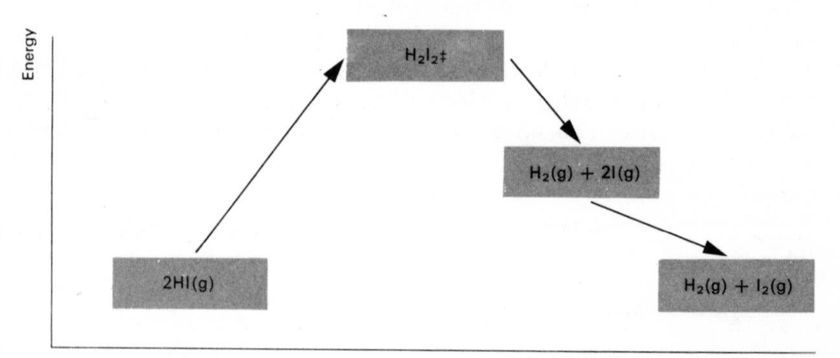

Figure 14.4b

The rate of reaction can be calculated from the rate at which an equilibrium between the reactants and the transition state is disturbed to give products.

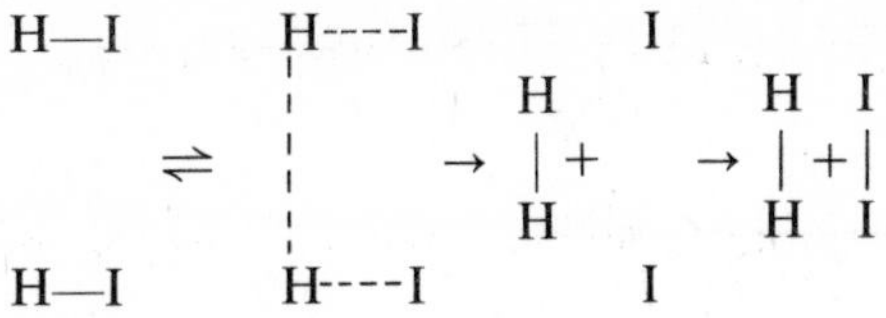

```
H—I         H----I            I
            ¦          H           H  I
      ⇌     ¦      →   | +     →   | + |
            ¦          H           H  I
H—I         H----I            I
```

Problem 6

The rate constant given by

$$\text{rate of disappearance of hydrogen iodide} = k[HI]^2$$

for the reaction $2HI(g) \rightarrow H_2(g) + I_2(g)$
varies with the temperature as in the following table.

Temperature T/K	**Rate constant** $k/10^{-5}\ dm^3\ mol^{-1}\ s^{-1}$
556	0.0352
647	8.58
700	116
781	3960

Plot a graph of $\lg k$ (on the vertical axis) against $\frac{1}{T}$ (on the horizontal axis) and from the gradient of this graph find the activation energy of the reaction.

Problem 7

The rate constant given by

$$\text{rate of appearance of nitrogen} = k\,[C_6H_5N_2Cl]$$

for the reaction

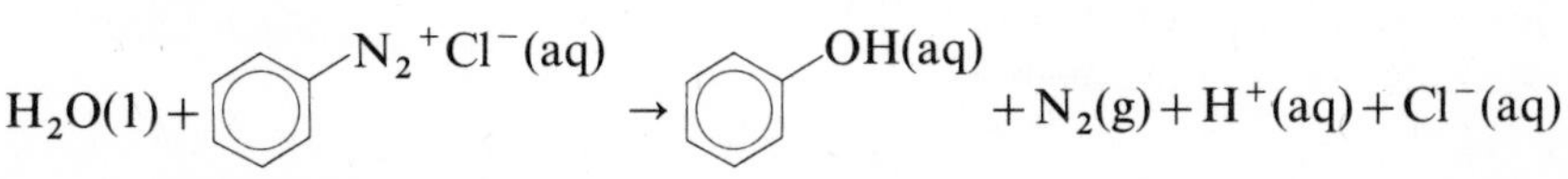

$$H_2O(l) + C_6H_5N_2^+Cl^-(aq) \rightarrow C_6H_5OH(aq) + N_2(g) + H^+(aq) + Cl^-(aq)$$

varies with temperature as in the following table.

Temperature T/K	**Rate constant** $k/10^{-5}\ s^{-1}$
278.0	0.15
298.0	4.1
308.2	20
323.0	140

Find the activation energy of the reaction.

14.5 An investigation of the kinetics of another reaction

Experiment 14.5

To investigate the kinetics of the reaction between permanganate and oxalate ions

In this experiment you are going to investigate the reaction between permanganate and oxalate ions, the equation for which is as follows:

$$2MnO_4^-(aq) + 16H^+(aq) + 5C_2O_4^{2-}(aq) \rightarrow 2Mn^{2+}(aq) + 8H_2O(l) + 10CO_2(g)$$

Because of the change of colour, the reaction is conveniently followed using a colorimeter.

Procedure

1 Using a burette, put 10 cm^3 of a solution which is 0.1M in oxalate ions and 1.2M in sulphuric acid into a test-tube.

2 Put 0.2 cm^3 of a 0.02M solution of permanganate ions into a test-tube that fits the colorimeter.

3 Adjust the meter of the colorimeter to maximum with a tube of water in place.

4 Add the oxalate solution to the permanganate solution, shake the mixture, and start the clock.

5 Put the tube of reaction mixture into the colorimeter and take readings every 20 seconds. As the reaction proceeds you may want to take readings more frequently, but as the experiment is completed in four to five minutes, it is easy to repeat it.

Convert the meter readings to molarities of permanganate (see below) and plot a graph of concentration of permanganate on the vertical axis against time on the horizontal axis. Compare it with the concentration of hydrogen peroxide against time graph in problem 2.

Try to answer the following questions:

1 How does the rate of the reaction change with time?

2 Can you suggest a reason for the changes in the rate that you observe?

3 Can you suggest any experimental work that you could do to see if your suggestion is correct?

Note on the use of the colorimeter

Use of a colorimeter is a quick and easy method of determining the concentrations of solutions which are coloured. Basically the instrument consists of the components shown in figure 14.5a.

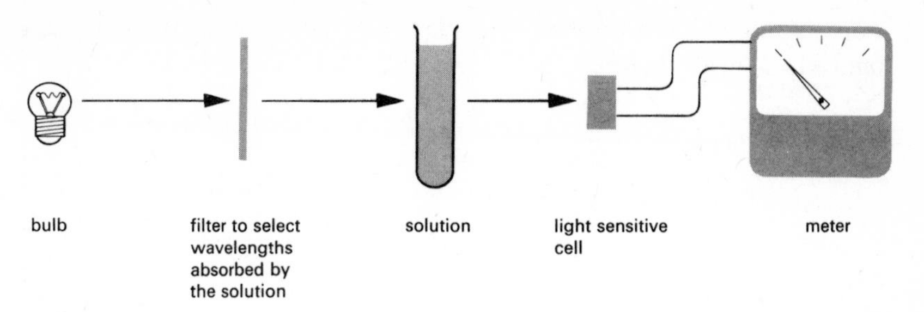

Figure 14.5a
A colorimeter.

The light-sensitive cell may either be a selenium cell which produces an e.m.f. proportional to the intensity of the light falling on it, or a cadmium sulphide cell, the electrical resistance of which is proportional to the intensity of the light falling on it. Either way the meter reading gives an indication of the intensity of light emerging from the solution.

The connection between the intensity of light emerging from the solution and the molarity of the absorbing species in the solution is

$$\lg\left(\frac{\mathrm{m_o}}{\mathrm{m}}\right) = \lg\left(\frac{I_o}{I}\right) \propto M$$

where I_o is the intensity of incident, monochromatic light, I is the intensity of emergent light, $\mathrm{m_o}$ and m are the meter readings, and M is the molarity of the solution.

The colorimeter is normally prepared for use by first adjusting the meter reading to maximum for I_o by inserting a tube of pure solvent and then adjusting the intensity of light by means of a shutter placed between the bulb and the cell. The tube of solution is then put in the colorimeter and a meter reading proportional to I determined. It is *most important* that this procedure is adopted for all readings if possible.

Unfortunately, the law quoted above is only obeyed accurately by certain solutions and under certain conditions. Also, the meter reading may not be accurately proportional to the intensity of light, so the instrument must be calibrated before use. This will also indicate for what range of concentration of a particular substance the colorimeter can be used.

For permanganate solutions in a particular colorimeter the curve shown in figure 14.5b was obtained.

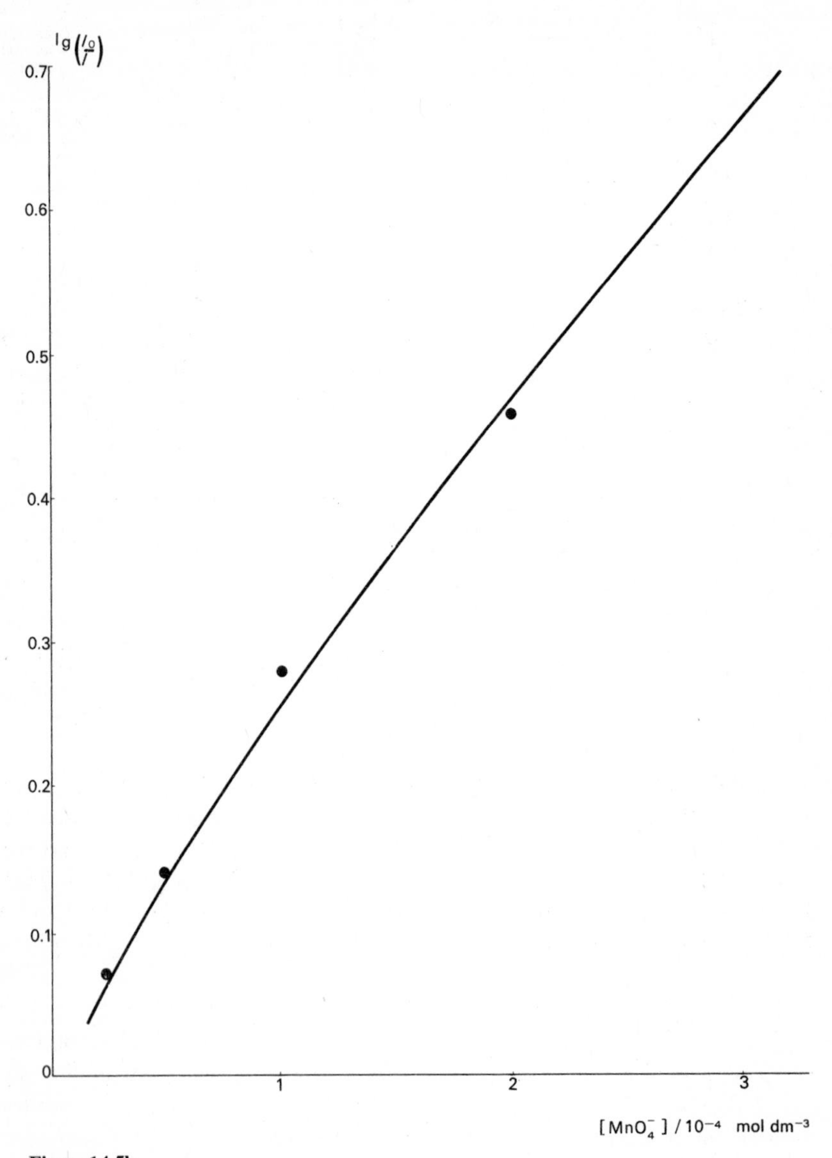

Figure 14.5b

If a calibration curve is being used, it is easier to plot $\left(\frac{I}{I_o}\right) = \left(\frac{m}{m_o}\right)$ against concentration than $\lg\left(\frac{I_o}{I}\right) = \lg\left(\frac{m_o}{m}\right)$, as m_o is normally 1, or 50 which are useful numbers to have as denominators. Such a graph (figure 14.5c) indicates that the colorimeter will not measure concentrations accurately above 0.0003M.

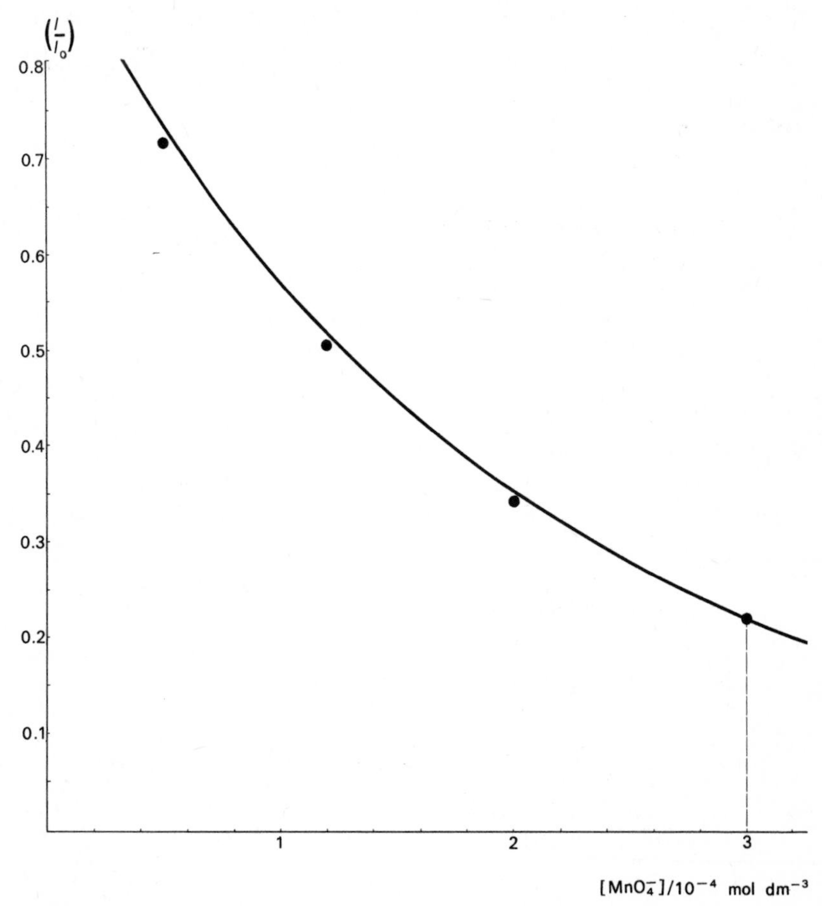

Figure 14.5c

Before these curves can be constructed, the most suitable filter must be chosen. This must select that band of wavelengths of light which are most strongly absorbed by the solution, that is, the filter which gives the lowest value of $\left(\frac{I}{I_o}\right)$ for a particular solution.

Procedure

You will probably be supplied with a calibration curve for a particular filter, colorimeter, and solution. You will require a test-tube for the pure solvent and an optically matched one for the solution. To take a reading, the meter is adjusted to maximum with pure solvent in place, then the reading is obtained with solution in place. It is best then to check again with solvent in place and obtain a meter reading with solution in place once again. The value of $\left(\frac{I}{I_o}\right)$ so obtained can then be turned into a molarity using the calibration chart.

Appendix

Integration of rate laws

For a first order reaction

A → products

$$-\frac{d[A]}{dt} = k_1 [A]$$

$$\therefore \quad -\frac{1}{[A]} d [A] = k_1 \, dt$$

Upon integration we have

$$-\int \frac{1}{[A]} d [A] = \int k_1 \, dt$$

$$\therefore \quad -\ln [A] = k_1 t + \text{constant} \qquad (1)$$

$$\left(\text{standard integral } \int \frac{1}{x} dx = \ln x\right)$$

At the start of the reaction when the concentration of A is $[A]_o$, $t = 0$, hence:

$$-\ln [A]_o = k_1 0 + \text{constant}$$

$$\therefore \quad -\ln [A]_o = \text{constant}$$

Substituting in (1);

$$-\ln [A] = k_1 t - \ln [A]_o$$

$$\therefore \quad \ln [A]_o - \ln [A] = k_1 t$$

$$\therefore \quad \ln \frac{[A]_o}{[A]} = k_1 t$$

$$\text{or} \quad \lg \frac{[A]_o}{[A]} = \frac{k_1}{2.303} t$$

Thus if a reaction is first order, a graph of $\lg \frac{[A]_o}{[A]}$ against t will be a straight line, and its gradient will be $\frac{k_1}{2.303}$. Note that absolute values of the concentrations are not required, only the ratio $\left(\frac{[A]_o}{[A]}\right)$, which is equal to $\left(\frac{V_o}{V}\right)$, where V_o and V are titres which are proportional to $[A]_o$ and $[A]$.

The relationship between the half-life of a first order reaction and the rate constant can be obtained by inserting the condition that when $t = t_{\frac{1}{2}}$, $[A] = \frac{[A]_o}{2}$, into the above equation:

$$\lg \left(\frac{[A]_0}{[A]_0/2}\right) = \frac{k_1}{2.303} t_{\frac{1}{2}}$$

$$\therefore \quad t_{\frac{1}{2}} = \frac{2.303 \lg 2}{k_1}$$

$$\therefore \quad t_{\frac{1}{2}} = \frac{0.69}{k_1}$$

indicating that half-life is constant and independent of the concentration at the beginning of each half-life.

For a second order reaction

$A + B \rightarrow$ products

(A and B may be the same:

$2A \rightarrow$ products)

where the initial concentration of A is equal to the initial concentration of B, that is $[A]_o = [B]_o$.

$$-\frac{d[A]}{dt} = k_2 [A]^2$$

$$\therefore \quad -\frac{1}{[A]^2} d[A] = k_2 dt$$

Upon integration we have

$$-\int \frac{1}{[A]^2} d[A] = \int k_2 dt$$

$$\therefore \quad \frac{1}{[A]} = k_2 t + \text{constant} \qquad (2)$$

$$\left(\text{standard integral } \int \frac{1}{x^2} dx = -\frac{1}{x}\right)$$

At the start of the reaction when the concentration of A is $[A]_o$ $(= [B]_o)$, $t = 0$, hence:

$$\frac{1}{[A]_o} = k_2 0 + \text{constant}$$

$$\therefore \quad \frac{1}{[A]_o} = \text{constant}$$

substituting in (2)

$$\frac{1}{[A]} = k_2 t + \frac{1}{[A]_o}$$

Thus if a reaction is second order, and the two initial concentrations are the same, a graph of $\frac{1}{[A]}$ against t will be a straight line, and its gradient will be k_2. Note that absolute values of [A] *are* required to obtain k_2, though titres, for instance, proportional to [A] would give a straight line indicating that a reaction is second order.

The relationship between half-life of a second order reaction and the velocity constant can be obtained by inserting the condition that when

$t = t_{\frac{1}{2}}$, $[A] = \frac{[A]_o}{2}$, into the above equation:

$$\frac{1}{[A]_o/2} = k_2 t_{\frac{1}{2}} + \frac{1}{[A]_o}$$

$$\therefore \quad t_{\frac{1}{2}} = \frac{1}{[A]_o k_2}$$

indicating that the half-life is not constant, depending on the concentration at the beginning of each half-life.

For higher order reactions

The formulae of obviously more complicated reactions can be worked out in a similar way to those above.

Topic 15

Equilibria: redox and acid-base systems

15.1 Redox equilibria: metal/metal ion systems

Earlier in the course, you have met a number of oxidation and reduction (redox) reactions and have seen, in Topic 5, that these always involve a change of oxidation number of the reacting substances. Redox reactions in which ions are involved will be studied in this and the next two sections. The principles governing these reactions provide a basis for understanding a very large number of chemical systems.

Experiment 15.1a

Some simple redox reactions

All the following reactions can be carried out in test-tubes.

1 Dip a strip of zinc foil into copper(II) sulphate solution (about 0.5M) and leave for about 30 seconds. Remove the strip and examine the zinc surface.

2 Repeat (1) using zinc powder instead of foil and find whether there is a temperature change during the reaction. Add zinc powder, with shaking, until there is no further change. Record any temperature change. Examine the solid and solution when reaction appears to be finished.

3 Repeat (1) using copper foil and silver nitrate solution (about 0.1M).

4 Repeat (2) using copper powder and silver nitrate solution.

Write ionic equations for what you have observed.
Was heat evolved or absorbed in the reactions?
Why do the changes in (2) and (4) proceed more quickly than those in (1) and (3)?
What can you say about the energy content of the products compared with the reactants for each system?
Write down the oxidation numbers of the reactants and the products in each reaction.
Which of the reactants has been oxidized and which reduced in each case?

Oxidation and reduction by electron transfer

Reactions between metals and metal ions involve the transfer of electrons from one reactant to another. This may be seen if the reactions are analysed into component reactions (or half-reactions), such as

$$Zn(s) \rightarrow Zn^{2+}(aq) + 2e^-$$

and $$2e^- + Cu^{2+}(aq) \rightarrow Cu(s)$$

In the first half-reaction, the oxidation state of zinc increases (0 to +2) so that an *oxidation* is involved; in the second reaction the oxidation state of copper decreases (+2 to 0) thus involving a *reduction*. In the complete reaction zinc is the *reductant* (or reducing agent) and copper the *oxidant* (or oxidizing agent). It will be seen that

loss of electrons is an *oxidation* process
gain of electrons is a *reduction* process
species that *gain electrons* are acting as *oxidants*
species that *lose electrons* are acting as *reductants*

Write half-reactions for the copper/silver nitrate reaction. Which species is the oxidant and which the reductant in this reaction?

From experiment 15.1a it will be seen that copper can act as either an oxidant or a reductant, depending on the conditions. This can be accounted for by treating each half-reaction as an equilibrium, so that we have

$$Cu(s) \rightleftharpoons Cu^{2+}(aq) + 2e^-$$

If electrons are added to this system, Le Chatelier's principle tells us that the equilibrium will move towards the left; removal of electrons will have the opposite effect. In the presence of a metal whose tendency to form ions in solution is greater than that of copper, the reaction moves towards the left and copper is deposited. This is the case with zinc, which dissolves to form hydrated zinc ions. With the silver system the reverse is the case and copper atoms lose electrons to become hydrated ions whilst metallic silver is precipitated.

Measuring the tendency of a metal to form ions in solution

If an equilibrium is set up when a metal is placed in an aqueous solution of its ions

$$M(s) \rightleftharpoons M^{z+}(aq) + ze^-$$

we should expect the metal to become negatively charged, by electrons building up on it, and the solution to become positively charged. Thus there should be an electric potential between solution and ions. If the equilibrium position differs for different metals, the potentials set up will differ also. These potentials (called absolute potentials) cannot be measured but the *difference* in potential between two metal/metal ion systems can be found by incorporating them into a voltaic cell and measuring the potential difference between the metal electrodes. Potential difference (p.d.) is measured in volts.

By using the circuit shown in figure 15.1a the variation of potential difference between metal electrodes in a voltaic cell with the resistance in the circuit can be explored.

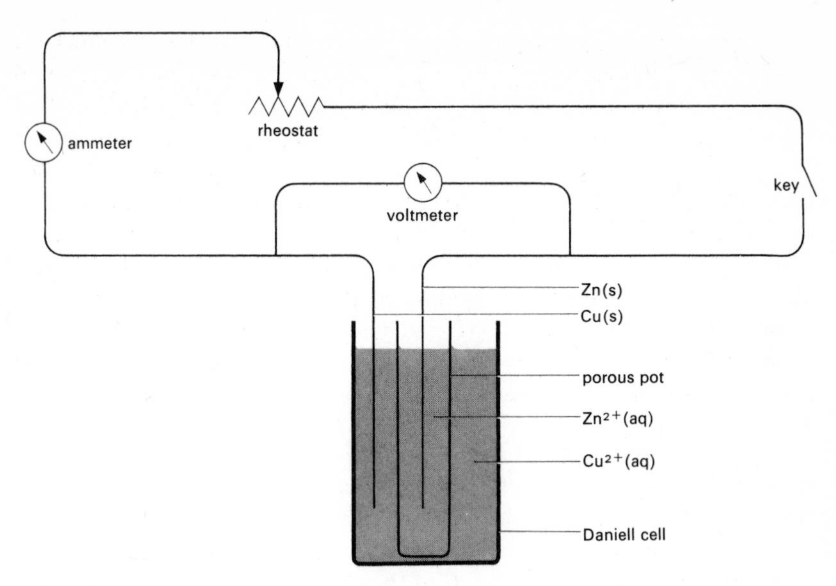

Figure 15.1a

In this example a Daniell cell is used. It transfers electrical energy to the circuit as a result of the reaction

$$Zn(s) + Cu^{2+}(aq) \rightarrow Zn^{2+}(aq) + Cu(s)$$

The Daniell cell is a combination of two electrode systems,

zinc metal immersed in zinc sulphate solution and
copper metal immersed in copper sulphate solution.

The solutions of the two systems are prevented from mixing by a porous pot (made of unglazed porcelain), the zinc sulphate solution usually being placed inside the pot and the copper sulphate solution outside. Electrical contact between the solutions is established in the walls of the porous pot.

When the cell is working, electrons are transferred from the zinc electrode to the external circuit

$$Zn(s) \rightarrow Zn^{2+}(aq) + 2e^-$$

and electrons are transferred from the circuit to the copper electrode

$$2e^- + Cu^{2+}(aq) \rightarrow Cu(s)$$

Thus the copper is the positive terminal of the cell, and the zinc the negative terminal.

Experiments in which the resistance in the circuit is varied, and using voltmeters of low and high resistance, show that the potential difference between the two metal electrodes increases as the resistance increases. The p.d. is at a maximum when no current is flowing. This maximum p.d. is called the electromotive force of the cell (e.m.f) which is denoted by the symbol E. When taking a measurement by a voltmeter (that is when the key is open in figure 15.1a) current must flow to operate the voltmeter. A direct voltmeter reading can therefore never give a highly accurate value for the e.m.f. of a cell. With a voltmeter of high resistance, such as a valve or transistor voltmeter, the current taken is extremely small, however, and only a minute error is involved in measurements of e.m.f. The value of E for a cell is a measure of the relative tendencies of the electrode systems involved to liberate electrons by forming ions in solution.

An alternative to the use of the valve voltmeter is the potentiometer method of measuring e.m.f. In this method, the current flowing in a circuit, of which the cell is a part, is opposed by a potential difference from another source of electricity, which can be adjusted until no current flows. The counter-voltage thus applied is equal to the e.m.f. of the cell being studied. A potentiometer is used to produce the counter-voltage (see appendix I at the end of this topic).

Cell diagrams

It is convenient to have an agreed method of representing voltaic cells and the e.m.f. which they produce. This is called a *cell diagram*. In Britain the convention agreed by the International Union of Pure and Applied Chemistry (IUPAC) is used for such diagrams. This may be illustrated by the Daniell cell, for which the cell diagram is written

$$Zn(s) \mid Zn^{2+}(aq) \,\vdots\, Cu^{2+}(aq) \mid Cu(s); \; E = +1.1 \text{ V}$$

The solid vertical lines represent boundaries between solids and solutions in each electrode system and the vertical broken line represents the porous partition (or other device to ensure a conducting path through the cell). The e.m.f. of the cell is represented by the symbol E, and the value is given in volts, with a sign (+ or −) preceding it which indicates the polarity of the *righthand electrode* in the diagram. In the example above the copper plate is the positive terminal of the cell. Obviously the cell diagram can be written in the reverse order but it is still the righthand electrode whose polarity is indicated. The Daniell cell can thus be written in the alternative form

$$Cu(s) \mid Cu^{2+}(aq) \,\vdots\, Zn^{2+}(aq) \mid Zn(s); \; E = -1.1 \text{ V}$$

This is the basic pattern for all cell diagrams. Additional conventions are required for more complicated cells. These will be dealt with as they arise.

Contributions made by separate electrode systems to the e.m.f. of a cell
Measurement of the potential of a single electrode system is impossible because two such systems are needed to make a complete cell of which the e.m.f. can be measured. We can, however, assess the *relative* contributions of single electrode systems to cell e.m.f.s by choosing one system as a standard against which all other systems are measured. The standard system is then arbitrarily assigned zero potential and the potentials of all other systems referred to this value. By international agreement the hydrogen electrode has been chosen as the reference electrode for this purpose.

The hydrogen electrode
From redox reactions such as

$$Mg(s) + 2H^+(aq) \rightleftharpoons Mg^{2+}(aq) + H_2(g)$$

it is clear that an equilibrium can be set up between hydrogen gas and its ions in solution

$$H_2(g) \rightleftharpoons 2H^+(aq) + 2e^-$$

By an arrangement such as that below this reaction can be used in a half-cell.

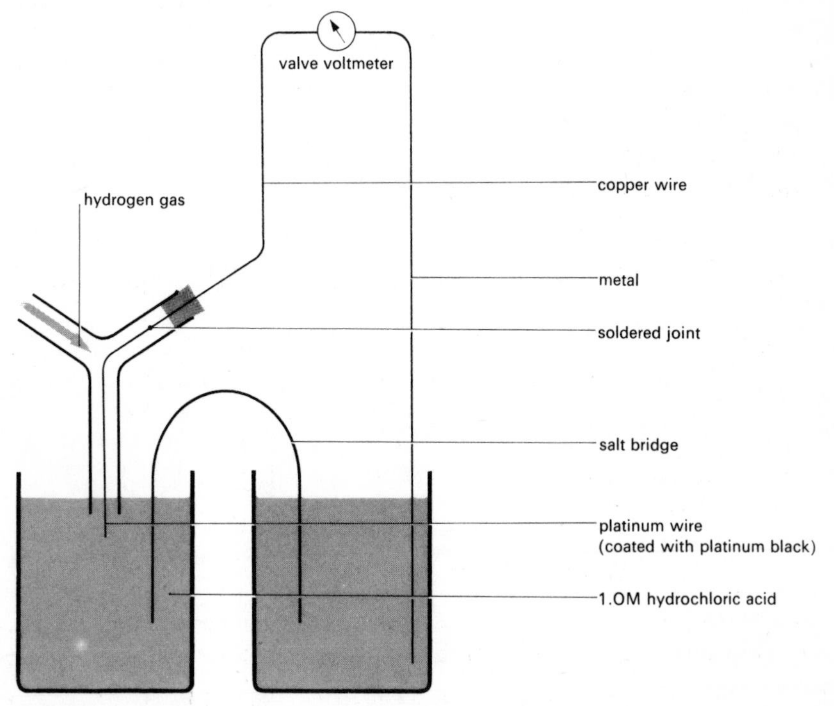

Figure 15.1b

This half-cell is called a *hydrogen electrode*. Essentially it consists of a platinum surface which is coated with finely divided platinum (usually called 'platinum black') which dips into a molar solution of hydrogen ions (usually 1 M HCl(aq)). A slow stream of pure hydrogen is bubbled over the platinum black surface. Under these conditions the equilibrium

$$H_2(g) \rightleftharpoons 2H^+(aq) + 2e^-$$

is established fairly quickly. The platinum black acts as a catalyst in this process and, being porous, it retains a comparatively large quantity of hydrogen. The platinum metal also serves as a convenient route by which electrons can leave or enter the electrode system. The hydrogen electrode is represented by

$$\text{Pt } [H_2(g)] \mid 2H^+(aq) \mid$$

By international agreement the hydrogen electrode is taken as the reference electrode. Its potential under specified conditions, detailed later, is taken as zero. The redox potential of any other metal is taken as the difference in potential between the metal electrode and the standard hydrogen electrode.

If the metal electrode is negative with respect to the hydrogen electrode, the standard redox potential is given a negative sign. If the metal electrode is positive with respect to the hydrogen electrode, the standard redox potential is given a positive sign.

Some values for *standard* redox potentials are given below. The conditions of temperature and concentration under which these are measured are specified later in this section (pages 74 and 75).

$Mg^{2+}(aq) \mid Mg(s)$	-2.37 V
$Zn^{2+}(aq) \mid Zn(s)$	-0.76 V
$Pb^{2+}(aq) \mid Pb(s)$	-0.13 V
$2H^+(aq) \mid [H_2(g)]$ Pt	0.00 V
$Cu^{2+}(aq) \mid Cu(s)$	$+0.34$ V
$Ag^+(aq) \mid Ag(s)$	$+0.80$ V

Note. Each of these values refers to the e.m.f. of a real cell

$$\text{Pt } [H_2(g)] \mid 2H^+(aq) \mid M^{z+}(aq) \mid M(s)$$

From the list, values for the e.m.f. of other cells can be estimated. What would be the value of the e.m.f., and the sign of the righthand electrode for the cell

$$Pb(s) \mid Pb^{2+}(aq) \mid Cu^{2+}(aq) \mid Cu(s) \ ?$$

As has been suggested earlier, a valve voltmeter (or a similar high resistance instrument) can give quite accurate values for cell e.m.f.s. An instrument of this kind gives a result much more quickly than a potentiometer. For this reason a valve voltmeter, if available, should be used for experimental work in the remainder of this topic where e.m.f. measurements are made.

Experiment 15.1d

To measure the e.m.f. of some voltaic cells

The circuit for this experiment is shown in figure 15.1c.

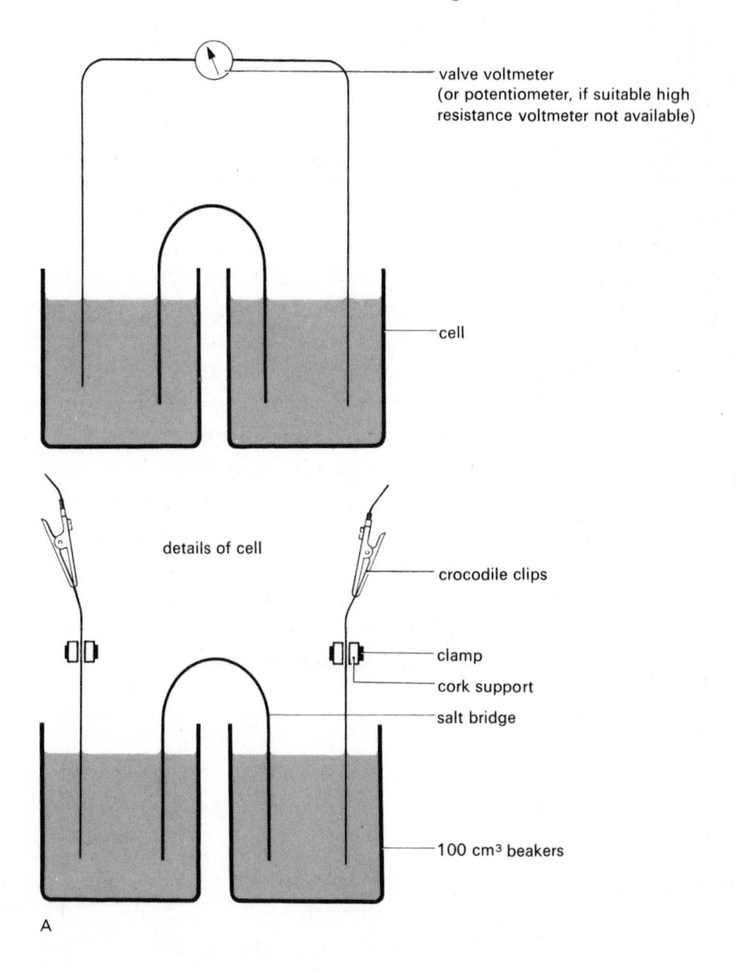

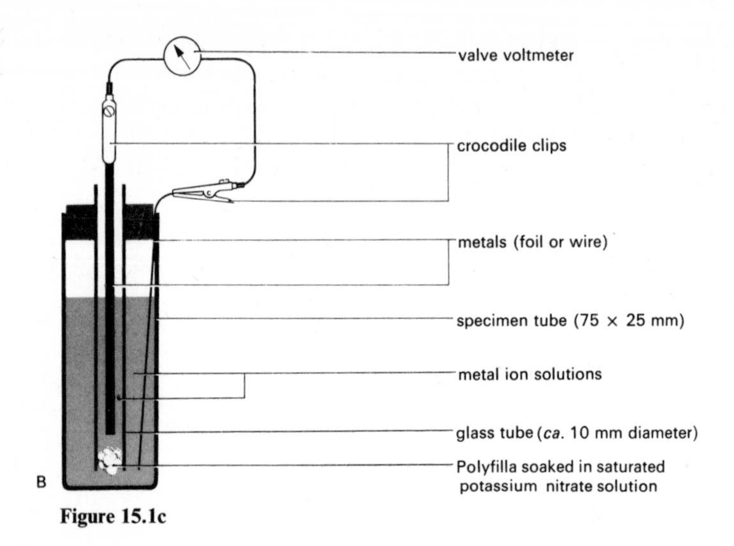

Figure 15.1c

Several types of cell can be used. That shown at A in figure 15.1c uses metal strips, approximately 6 cm by 1 cm, as electrodes; they are slotted into pieces of cork held in small clamps, so that they dip into electrode solutions contained in 100 cm^3 beakers. The metal strips are cleaned with emery paper before use. Each beaker contains a 'half-cell' and connection between them is made by a single strip of filter paper, approximately 10 cm × 1 cm, soaked in saturated potassium nitrate solution (the 'salt bridge'). Connections to the metal electrodes are made with crocodile clips. Alternative B is more portable and uses smaller volumes of solutions. The salt bridge is a plug of Polyfilla soaked in saturated potassium nitrate solution. A further variation can be made by using a 100 × 12 mm test-tube, with a hole blown in the rounded part, instead of the inner tube in version B. Small pieces of filter paper, soaked in saturated potassium nitrate solution and compressed into a plug about 1 cm thick (use a glass rod) form the salt bridge.

Method

Set up the cell

$$Cu(s) \mid Cu^{2+}(aq) \vdots Zn^{2+}(aq) \mid Zn(s) \qquad \text{(a)}$$

using 1.0M $ZnSO_4(aq)$ and 1.0M $CuSO_4(aq)$ as the electrode liquids. Connect the cell to the valve voltmeter and measure the e.m.f. You should be able to forecast which is the positive pole of this cell, and make the voltmeter connections accordingly.

Repeat the measurements for the following cells, using a fresh salt bridge each time a half-cell is changed

$$Ag(s) \mid Ag^{+}(aq) \vdots Cu^{2+}(aq) \mid Cu(s) \qquad \text{(b)}$$

$$Ag(s) \mid Ag^{+}(aq) \vdots Zn^{2+}(aq) \mid Zn(s) \qquad \text{(c)}$$

For the silver half-cell use 0.1M $AgNO_3$ with silver foil; a small piece of copper foil as backing, held in the cork but well clear of the solution, will give a more permanent contact for attaching the crocodile clip, if you are using cell A.

Questions

1 Try to relate the results of this experiment to those of the displacement reactions in experiment 15.1a. Is the system with the greatest tendency to form ions the positive or negative pole of each cell?

2 Is the e.m.f. for cell (c) what you would expect from the values obtained for cells (a) and (b)?

You could obtain the equivalent of cell (c) by connecting cells (a) and (b) together as follows

$$Ag(s) \mid Ag^{+}(aq) \vdots Cu^{2+}(aq) \mid \overbrace{Cu(s) \quad Cu(s)} \mid Cu^{2+}(aq) \vdots Zn^{2+}(aq) \mid Zn(s)$$

The two copper electrodes would then cancel each other.

3 How do the results obtained for each cell compare with the e.m.f. values calculated from the list of standard redox potentials given on page 70?

Experiment 15.1e

To investigate the effect of changes in silver ion concentration on the potential of the $Ag^{+}(aq) \mid Ag(s)$ electrode

By applying the principle of Le Chatelier to the system

$$Ag(s) \rightleftharpoons Ag^{+}(aq) + e^{-}$$

work out the effect of altering the ionic concentration of the solution. If the solution is made more dilute, will the metal tend to form ions to a greater or lesser extent? Will this result in the electrode potential becoming more negative or more positive?

Test your predictions by measuring the e.m.f. of the cell

$$Cu(s) \mid Cu^{2+}(aq) \vdots Ag^{+}(aq) \mid Ag(s)$$

keeping the ion concentration in the copper electrode system constant and varying the ion concentration in the silver electrode system.

Procedure

The same circuit as that for experiment 15.1d can be used; the valve voltmeter can be shared on a communal basis. If a valve voltmeter, or other suitable high resistance voltmeter, is not available, a potentiometer circuit can be used.

Measure the e.m.f. of the cell for various values of $[Ag^+(aq)]$, keeping the concentration of copper ion constant at 1.0M in the other electrode. Suitable silver ion concentrations are 0.01M, 0.0033M, 0.001M, 0.00033M, and 0.0001M. The concentrations other than 0.01M can be obtained by progressive dilution of 0.01M solution. Extreme care and attention to cleanliness are essential when handling very dilute solutions. A fresh salt bridge (saturated potassium nitrate solution) must be used for each new concentration of silver ions. If a cell of the type shown at B in figure 15.1c is used, each group of students could make up one cell. The cells could then be passed round so that each group measures the e.m.f. of every cell in the range quoted above.

Record the results in tabular form, using separate columns for $[Ag^+(aq)]/\text{mol dm}^{-3}$, $\lg [Ag^+(aq)]$, and E/volt.

Plot a graph of E against $\lg [Ag^+(aq)]$, with E values on the vertical axis and $\lg [Ag^+(aq)]$ values on the horizontal axis. Choose a horizontal scale so that extrapolation to $\lg [Ag^+(aq)] = -18$ is possible. This graph will be needed in a later experiment (experiment 15.2a).

Questions

1 Does e.m.f variation with ion concentration follow the predictions made by applying Le Chatelier's principle to the $Ag(s)|Ag^+(aq)$ system?

2 From the shape of the graph what can be deduced about the relation between electrode potential and ion concentration?

3 What concentration of silver ion would make the silver electrode potential exactly equal to the copper electrode potential for 1.0M $Cu^{2+}(aq)$? The cell e.m.f. will then be zero. Under these conditions the two electrodes will be in equilibrium, as represented by the equation

$$Cu(s) + 2Ag^+(aq) \rightleftharpoons Cu^{2+}(aq) + 2Ag(s)$$

Calculate the equilibrium constant, K_c, for this reaction.

Standard electrode potentials

From experiment 15.1e it can be seen that the value of E for a given electrode system varies with the ionic concentration of the solution. It can also be shown that E varies with the temperature. Thus standardized conditions of temperature and concentration must be used if electrode potential measurements are to be compared. By international agreement the conditions chosen are:

an ion concentration of one mole per cubic decimetre*
a temperature of 298 K (25 °C)

The value of an electrode potential relative to the *standard hydrogen electrode* under these conditions is called the *standard redox potential* (or standard electrode potential). By definition again, the standard hydrogen electrode consists of hydrogen gas at one atmosphere pressure bubbling over platinized platinum in a solution of hydrogen ion concentration one mole per cubic decimetre.* Standard redox potentials are denoted by the symbol $E^{\ominus}$. A series of $E^{\ominus}$ values is given in the *Book of Data* (table TEI).

For other ion concentrations at around room temperature, the value of E can be calculated approximately from the expression

$$E = E^{\ominus} + \frac{0.06}{z} \lg [\text{ion}]$$

where z = the charge on the metal ion.

This expression applies to metal/metal ion electrodes *only*; other electrode systems behave differently, and will be dealt with later. The value of the constant (0.06) varies slightly with temperature; it is 0.057 at 15 °C, 0.059 at 25 °C, and 0.060 at 30 °C. It will be seen that if [ion] is less than one, the second term on the righthand side is negative since lg [ion] will then be negative. Hence E becomes less positive, or more negative, with dilution.

The equation

$$E = E^{\ominus} + \frac{0.06}{z} \lg [\text{ion}]$$

has the general form

$$y = \text{c} + \text{m}x$$

* Owing to incomplete ionization, interference between ions, and effects of 'crowding together' in fairly concentrated solution, the actual concentration of dissolved material is usually greater than this. Thus to obtain a concentration of free hydrogen ions of one mole per cubic decimetre a solution of hydrogen chloride which is 1.18M must be used. We shall ignore these effects in the simplified treatment given in this topic.

If y is plotted against x, a straight line is obtained. The intercept on the y axis for $x = 0$ gives the value of c, and the slope of the line (calculated in terms of the units represented on the axes) gives the value of m. For our purposes, E is y, lg [ion] is x, and $\frac{0.06}{z}$ is m.

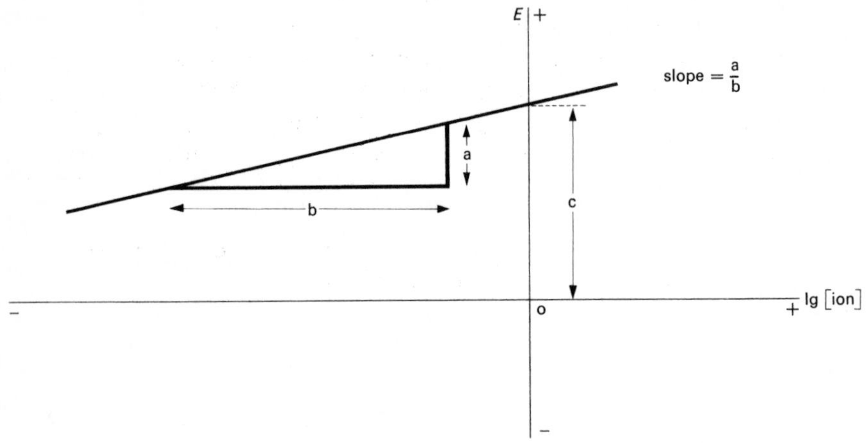

Figure 15.1d

You can see whether the results obtained in experiment 15.1e fit into the above expression for calculating E. This is best done by calculating the values of E on the hydrogen scale for the electrode in which the concentration was varied. In experiment 15.1e this was the silver electrode. Thus we have

$$E_{cell} = E_{Ag} - E^{\ominus}_{Cu}$$

E_{cell} = e.m.f. of cell at various values of $[Ag^+(aq)]$
E_{Ag} = redox potential of silver electrode
$E^{\ominus}_{Cu}$ = standard redox potential of copper electrode (1.0M solution was used)

Rearranging the equation above,
we have $E_{Ag} = E_{cell} + E^{\ominus}_{Cu}$

From the *Book of Data,* table TEI, the value of $E^{\ominus}_{Cu}$ is 0.34 V.

So $E_{Ag} = E_{cell} + 0.34$ V.

Calculate the values of E_{Ag} for the various concentrations of silver ion used in the experiment. Plot these values against lg $[Ag^+(aq)]$, using the vertical axis for E values and the horizontal axis for lg $[Ag^+(aq)]$. Draw the best straight line through the points plotted and calculate the slope of this line in terms of the units represented on the axis. The charge on the silver ion is +1, so the slope should be 0.06 if your results fit with the equation

$$E = E^{\ominus} + \frac{0.06}{z} \lg [\text{ion}]$$

The relationship between E and [ion] arises from the *Nernst equation.*

$$E = E^{\ominus} + \frac{RT}{zF} \ln [\text{ion}]$$

where R is the gas constant (the value of 8.313 J K^{-1} must be used)
T is the temperature in K
F is the faraday (96 500 coulombs)
z is the number of positive charges on the metal ion
ln [ion] is the logarithm of the ion concentration, in mol dm^{-3}, to the base e (2.718)

To convert logarithms to the base 10 (lg) to logarithms to the base e (ln) multiply by 2.3. The Nernst equation then becomes

$$E = E^{\ominus} + \frac{2.3RT}{zF} \lg [\text{ion}]$$

The Nernst equation cannot be used to calculate the value of E at different temperatures since $E^{\ominus}$ varies with temperature also. The potential of the standard hydrogen electrode, by definition, is zero at all temperatures.

By making use of known variation of ion concentration with E, the concentration of ions in very dilute solutions can be measured electrically. This method has drawbacks with concentrated solutions but can be used for solutions which would be much too dilute to be analysed by ordinary chemical methods. Some examples of the method are contained in the next section.

15.2 Use of e.m.f. measurements to estimate small concentrations of ions

Experiment 15.2a

To investigate the changes in silver ion concentration in solution when different reagents are added

The cell

$$Cu(s) \mid Cu^{2+}(aq) \,\vdots\, Ag^{+}(aq) \mid Ag(s)$$

is set up, using a standard copper electrode with 1.0M Cu^{2+}(aq). The silver electrode is made up using different mixtures of silver nitrate and other reagents. From e.m.f. measurements made on the cells which result, the silver ion concentrations in the mixtures can be found by using the graph prepared from the results of experiment 15.1e.

Procedure

This is the same as for experiment 15.1e.

Investigate as many of the following systems as you can. *For some of the measurements you may have to reverse the connections to the test cell. Use laboratory time for making measurements; the calculations can be done later.*

1 Mix one volume of 0.1M silver nitrate solution with two volumes of 0.1M potassium chloride solution. The actual volumes taken will depend on the type of cell that you are using. Insert a strip of silver foil into the mixture and measure the e.m.f. of the cell formed by this electrode and the standard copper electrode.

Calculation

Use the first graph plotted from the results of experiment 15.1e to calculate the approximate silver ion concentration in the solution of the silver electrode. This gives the value of $[Ag^{+}(aq)]_{eqm}$ in

$$AgCl(s) \rightleftharpoons Ag^{+}(aq) + Cl^{-}(aq)$$

The approximate value for the chloride ion concentration in this equilibrium can be obtained by assuming that all the silver ions used originally react with chloride ions; this is very nearly true since silver chloride is very sparingly soluble. Thus half the chloride ions added (those in one of the two volumes of potassium chloride solution used) are removed by precipitation. Since there is a total of three volumes of mixture,

$$[Cl^{-}(aq)]_{eqm} = \tfrac{1}{3} \times 0.1 \text{ mol dm}^{-3}$$

Calculate the approximate value of the solubility product of silver chloride

$$K_{sp} = [Ag^{+}(aq)]_{eqm}[Cl^{-}(aq)]_{eqm}$$

Compare your result with the value given in the table of solubility products in the *Book of Data.*

Repeat the procedure in (1) using the following mixtures:

2 1 volume 0.1M silver nitrate + 2 volumes 0.1M potassium bromide
3 1 volume 0.1M silver nitrate + 2 volumes 0.1M potassium iodide
4 1 volume 0.1M silver nitrate + 2 volumes 0.1M potassium iodate

Does the value of K_{sp} obtained for silver iodate agree with your results for experiment 12.3? If not, can you account for the difference?

Solubility products and solubility

An approximate value for the solubility of a substance can be calculated if its solubility product is known, and vice versa. The values obtained can only be taken as approximate since considerable errors are involved in measurement of both solubility product and solubility for the sparingly soluble substances to which the simple equilibrium law can be applied. Two examples will show how the calculations can be done.

Example 1

The solubility product of lead(II) chloride is 2×10^{-5} mol^3 dm^{-9} at 25 °C. What is the approximate solubility of the salt at this temperature?
The equilibrium involved is

$$PbCl_2(s) \rightleftharpoons Pb^{2+}(aq) + 2Cl^-(aq)$$

∴ For a saturated solution of lead chloride there will be two moles of chloride ions for each mole of lead ions.

In symbols, $[Cl^-(aq)]_{eqm} = 2 \times [Pb^{2+}(aq)]_{eqm}$

Now, $K_{sp} = [Pb^{2+}(aq)]_{eqm}[Cl^-(aq)]^2_{eqm}$
$= 4 \times [Pb^{2+}(aq)]^3_{eqm}$

So $$[Pb^{2+}(aq)]_{eqm} = \sqrt[3]{\frac{K_{sp}}{4}} = \sqrt[3]{\frac{2 \times 10^{-5}}{4}} \text{ mol dm}^{-3}$$

$$= 1.7 \times 10^{-2} \text{ mol dm}^{-3} \text{ at } 25\,^\circ\text{C}$$

One mole of Pb^{2+}(aq) ions is obtained from one mole $PbCl_2$
∴ 1.7×10^{-2} mole Pb^{2+}(aq) ions is obtained from 1.7×10^{-2} mole $PbCl_2$
One mole of $PbCl_2$ weighs $207 + 2 \times 35.5 = 278$ g
∴ One cubic decimetre of solution contains $1.7 \times 10^{-2} \times 278$
$= 4.8$ g $PbCl_2$
∴ Approximate solubility of lead chloride at 25 °C is 4.8 g dm^{-3}

Example 2

The solubility of silver chromate at 25 °C is 3.2×10^{-2} g dm^{-3}. What is its approximate solubility product at this temperature?

One mole of Ag_2CrO_4 contains $2 \times 108 + 52 + 4 \times 16$ g
$= 332$ g

∴ 3.2×10^{-2} g Ag_2CrO_4 is $\dfrac{3.2 \times 10^{-2}}{332} = 9.6 \times 10^{-5}$ mole

∴ The saturated solution of Ag_2CrO_4 is $2 \times 9.6 \times 10^{-5}$M with respect to Ag^+(aq) ions and 9.6×10^{-5}M with respect to CrO_4^{2-}(aq) ions.

$\therefore\ [Ag^{+}(aq)]_{eqm} = 2 \times 9.6 \times 10^{-5}\ mol\ dm^{-3}$

and $[CrO_4^{2-}(aq)]_{eqm} = 9.6 \times 10^{-5}\ mol\ dm^{-3}$

The equilibrium involved is

$$Ag_2CrO_4(s) \rightleftharpoons 2Ag^{+}(aq) + CrO_4^{2-}(aq)$$

$$\begin{aligned} K_{sp} &= [Ag^{+}(aq)]^2_{eqm}[CrO_4^{2-}(aq)]_{eqm} \\ &= (2 \times 9.6 \times 10^{-5})^2\,(9.6 \times 10^{-5})\ mol^3\ dm^{-9} \\ &= 4\,(9.6 \times 10^{-5})^3\ mol^3\ dm^{-9} \\ &= 3.5 \times 10^{-12}\ mol^3\ dm^{-9} \text{ at } 25\ ^{\circ}C \end{aligned}$$

Experiment 15.2b

A revision exercise

Use information obtained in experiment 15.2a and from the *Book of Data* to predict what will happen when

1 Silver nitrate solution is added drop by drop to a solution containing chloride and iodide ions

2 Solid silver chloride is shaken with potassium iodide solution

3 Solid silver chromate is shaken with potassium chloride solution

4 Solid silver iodate is shaken with potassium bromide solution.

Carry out test-tube experiments to test your predictions. Make the insoluble silver salts by precipitation from silver nitrate solution and decant off surplus liquid. Use small quantities of precipitate.

15.3 Redox equilibria extended to other systems

So far in this topic we have been mainly concerned with redox systems involving a metal in equilibrium with its ions in solution. Earlier in the course, however, we have encountered redox systems of other kinds in which reactions between non-metals and non-metal ions, and between ions only, take place. Examples of these are

$$2Br^{-}(aq) \rightleftharpoons Br_2(aq) + 2e^{-}$$

and $MnO_4^{-}(aq) + 8H^{+}(aq) + 5e^{-} \rightleftharpoons Mn^{2+}(aq) + 4H_2O(l)$

We shall now see how reactions of this kind can be fitted into the pattern of redox potentials which we have developed in sections 15.1 and 15.2. To begin with, we shall investigate a reaction in which there is no solid metal involved.

Experiment 15.3a

To investigate the reaction between iron(III) ions and iodide ions

To about 2 cm^3 of a solution which is approximately 0.1M with respect to Fe^{3+}(aq) ions – iron(III) sulphate or ammonium iron(III) sulphate (iron alum) is suitable – add an equal volume of approximately 0.1M potassium iodide solution.

Test separate portions of the original solutions and the final mixture with

1 starch solution,

2 potassium hexacyanoferrate(III) solution (sometimes known as potassium ferricyanide solution).

Add potassium hexacyanoferrate(III) solution to a solution containing Fe^{2+}(aq) ions – iron(II) sulphate is suitable.

The course of the reaction should be obvious from these tests, and from the colour change when the solutions are mixed. Write an equation for the reaction.

Using other redox reactions in voltaic cells

The equation for the reaction between iron(III) ions and iodide ions, studied in experiment 15.3a is

$$2Fe^{3+}(aq) + 2I^-(aq) \rightarrow 2Fe^{2+}(aq) + I_2(aq)$$

Each element has undergone a change of oxidation number for which we can postulate two separate processes

$Fe^{3+}(aq) + e^- \rightarrow Fe^{2+}(aq)$ *reduction*

and $2I^-(aq) \rightarrow I_2(aq) + 2e^-$ *oxidation*

By analogy with half-reactions studied earlier we might expect two competing equilibria:

$$Fe^{3+}(aq) + e^- \rightleftharpoons Fe^{2+}(aq)$$
$$2I^-(aq) \rightleftharpoons I_2(aq) + 2e^-$$

with equilibrium positions which differ for each reaction. When the two systems are brought together the tendency for the iodide/iodine system to liberate electrons is greater than that of the iron(II)/iron(III) system. Thus the first equilibrium moves to the right and Fe^{2+}(aq) ions are formed. Loss of electrons from the second equilibrium causes this to move to the right. This process will continue until the two systems are in equilibrium with each other. The equilibrium position for the overall reaction

$$2Fe^{3+}(aq) + 2I^-(aq) \rightleftharpoons 2Fe^{2+}(aq) + I_2(aq)$$

lies well over to the righthand side ($K_c \approx 10^7$ at 25 °C) so that we can say it is virtually complete.

It should be possible to use this reaction in a voltaic cell. The only problem is that of providing a means of allowing electrons to leave and enter the electrode systems. The solution is to use an inert electrode, as was done for the hydrogen half-cell in section 15.1, but in this case the electrode need not function as a catalyst, so smooth platinum is suitable. The following electrode systems can therefore be set up and their *E* values determined using a reference electrode

$Fe^{3+}(aq), Fe^{2+}(aq) \mid Pt$

and $I_2(aq), 2I^-(aq) \mid Pt$

We need to extend our conventions for writing cell diagrams in order to deal with systems such as these. The accepted practice is to put the *reduced* form (that in which the oxidation number is lowest) of the electrode system nearest to the electrode, and separate it from the oxidized form by a comma. (In the above examples the oxidation numbers are Fe^{3+}, +3; Fe^{2+}, +2; I_2, 0; I^-, −1.) Since iodine dissolves to a very small extent only in water and is usually obtained in solution in aqueous potassium iodide, it is sometimes represented as $I_2/KI(aq)$ in cell diagrams.

Experiment 15.3b

To measure the redox potentials for the $Fe^{3+}(aq)/Fe^{2+}(aq)$ equilibrium and the $2I^-(aq)/I_2(aq)$ equilibrium

The circuit used is shown in figure 15.3.

Procedure

In order to allow the half-cell reactions to proceed in either direction, we must have both the oxidized and reduced forms present in the electrode systems. The lefthand electrode in figure 15.3 must therefore contain a solution in which both $Fe^{3+}(aq)$ ions and $Fe^{2+}(aq)$ ions are present for the first measurement, and $I^-(aq)$ ions and I_2 molecules for the second measurement. Keep the beakers containing these solutions for the third e.m.f. measurement. The $Cu(s) \mid Cu^{2+}(aq)$ system is used as a reference electrode.

Set up the following cells and measure their e.m.f.

$Cu(s) \mid Cu^{2+}(aq) \,\vdots\, Fe^{3+}(aq), Fe^{2+}(aq) \mid Pt$

and $Cu(s) \mid Cu^{2+}(aq) \,\vdots\, I_2(aq), 2I^-(aq) \mid Pt$

From the results calculate the e.m.f. of the cell

$Pt \mid 2I^-(aq), I_2(aq) \,\vdots\, Fe^{3+}(aq), Fe^{2+}(aq) \mid Pt$

Check your calculation by measuring the e.m.f. of this cell. A second platinum electrode will be needed for this.

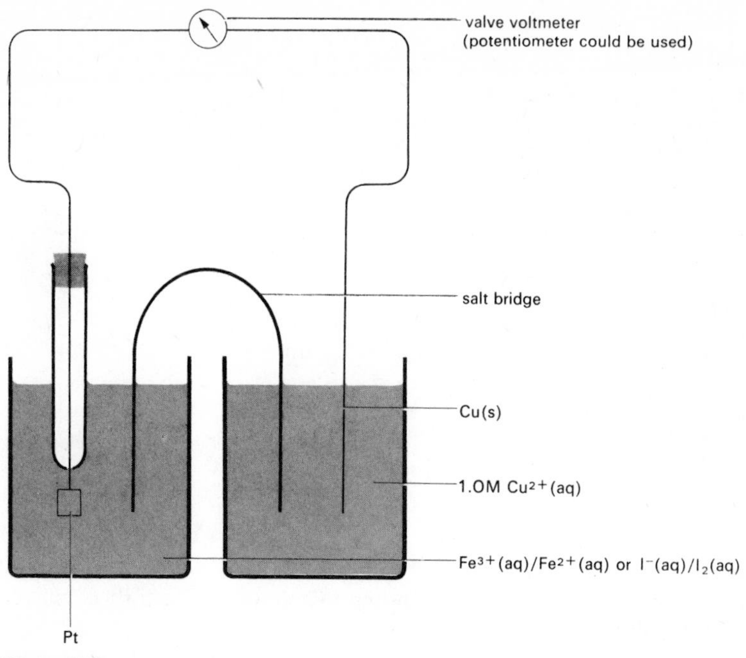

Figure 15.3

Concentration effects in ion/ion systems

From the results of experiment 15.3b it will be seen that ion/ion systems and non-metal/non-metal ion systems can be used in cell reactions in exactly the same way as metal/metal ion systems.

The application of either the principle of Le Chatelier or the equilibrium law to the iron(II)/iron(III) equilibrium indicates that the equilibrium position is affected by relative ion concentrations.

The greater the relative concentration of Fe^{3+}(aq) ions the more the equilibrium

$$Fe^{3+}(aq) + e^- \rightleftharpoons Fe^{2+}(aq)$$

moves to the right. This will *reduce* the absolute negative potential of the system, so that measured against the copper or hydrogen reference electrodes the difference in potential will *increase*, that is, it will become more positive.

The concentration effect is shown quantitatively by the figures given in table 15.3.

Relative concentrations /mol dm^{-3}		$\lg \frac{[Fe^{3+}]}{[Fe^{2+}]}$	E/volt
$[Fe^{3+}(aq)]$	$[Fe^{2+}(aq)]$		
1	9	−0.954	0.716
2	8	−0.602	0.735
3	7	−0.368	0.748
4	6	−0.177	0.760
5	5	0	0.770
6	4	+0.177	0.782
7	3	+0.368	0.792
8	2	+0.602	0.805
9	1	+0.954	0.825

Table 15.3
The variation of redox potential with concentration for the $Fe^{3+}(aq)$, $Fe^{2+}(aq)$ electrode (E values measured against a standard hydrogen electrode)

As with metal/metal ion systems, temperature also has an effect on ion/ion equilibria. It is therefore necessary to specify both concentration and temperature conditions for standard redox potentials involving equilibria between ions. The conditions chosen are:

equal molar concentrations of the reduced and oxidized forms of ion
a temperature of 298 K (25 °C)

The symbol $E^\ominus$ is again used to indicate a standard redox potential.

The value of E for other conditions is given by the Nernst equation in the form

$$E = E^\ominus + \frac{RT}{zF} \ln \frac{[\text{oxidized form}]}{[\text{reduced form}]}$$

which gives, for a temperature of 25 °C,

$$E = E^\ominus + \frac{0.06}{z} \lg \frac{[\text{oxidized form}]}{[\text{reduced form}]}$$

For ion/ion systems z is the number of electrons transferred when the oxidized form changes to the reduced form. For the equilibrium

$$\underset{\text{(oxidized form)}}{Fe^{3+}(aq)} + e^- \rightleftharpoons \underset{\text{(reduced form)}}{Fe^{2+}(aq)}$$

$z = 1$.

If the values of $\lg \frac{[Fe^{3+}(aq)]}{[Fe^{2+}(aq)]}$ are plotted against E values from table 15.3, a straight line is obtained of slope 0.06. For this system, the standard redox potential (value of E when $[Fe^{3+}(aq)] = [Fe^{2+}(aq)]$ and hence $\frac{[Fe^{3+}(aq)]}{[Fe^{2+}(aq)]} = 1$) is 0.770 V.

The form of the Nernst equation used earlier for metal/metal ion electrodes

$$E = E^{\ominus} + \frac{0.06}{z} \lg [\text{ion}]$$

is a special case of the general redox equation given above. For metal/metal ion systems the ion is the oxidized form and the metal the reduced form. Since the metal is a solid its concentration is constant and is thus not included in the equation. For non-metal/non-metal ion systems the non-metallic element is the oxidized form and the ion the reduced form.

Some further notes on standard redox potentials

You should now be in a position to appreciate all the information given in table TEI of the *Book of Data* but the following notes may be helpful in using this and other similar tables of $E^{\ominus}$ values.

1 It sometimes happens that the reduced and oxidized parts of an electrode system contain more than one chemical species (ion or molecule) which take part in the cell reaction. For example, the permanganate ion generally exerts its oxidizing power in presence of hydrogen ions, and water molecules are formed amongst the products of oxidation. These ions and molecules must be included in the oxidized and reduced forms of the equilibrium mixture.

$$MnO_4^-(aq) + 8H^+(aq) + 5e^- \rightleftharpoons Mn^{2+}(aq) + 4H_2O(l)$$

The half-cell diagram for this system is written

$$[MnO_4^-(aq) + 8H^+(aq)], [Mn^{2+}(aq) + 4H_2O(l)] \mid Pt$$

The square brackets in this and similar diagrams do not stand for 'the concentration of' but are merely used to bracket together the oxidized and reduced forms of the equilibrium mixture.

The Nernst equation for calculating E values at other than standard concentration conditions is

$$E = E^{\ominus} + \frac{0.06}{z} \lg \frac{[MnO_4^-(aq)]\,[H^+(aq)]^8}{[Mn^{2+}(aq)]}$$

In this equation, the square brackets have their usual significance. The concentration of water is not included since the variation of this is negligible in aqueous solutions. It will be obvious that E values for this and similar electrode systems are very sensitive to changes in hydrogen ion concentration.

2 In some tables of redox potentials the electrode systems are given in the form

$$M^{z+}(aq) + ze^- = M(s);\ E^\ominus = \pm x\ V$$

Examples are

$$Cu^{2+}(aq) + 2e^- = Cu(s);\ E^\ominus = +0.34\ V$$
$$Zn^{2+}(aq) + 2e^- = Zn(s);\ E^\ominus = -0.76\ V$$
$$IO_3^-(aq) + 6H^+(aq) + 5e^- = \tfrac{1}{2}I_2(aq) + 3H_2O(l);\ E^\ominus = +1.19\ V$$

It is easy to convert these to

$$Cu^{2+}(aq)\,|\,Cu(s)$$
$$Zn^{2+}(aq)\,|\,Zn(s)$$
$$[IO_3^-(aq) + 6H^+(aq)],\ [\tfrac{1}{2}I_2(g) + 3H_2O(l)]\,|\,Pt, \text{ if required.}$$

Note. In some older textbooks, everything is written the other way round, and all the signs are reversed, e.g. $Cu(s) = Cu^{2+}(aq) + 2e^-;\ E^\ominus = -0.34$ V.

3 Not all the $E^\ominus$ values given in tables have been obtained by direct measurements. Many of them are calculated from other experimental data.

Some uses of $E^\ominus$ values

Two important uses which can be made of tabulated $E^\ominus$ values are:

1 *Calculating the e.m.f. of voltaic cells* – The procedure for this has been mentioned earlier. Write the cell diagram, reverse the sign of the $E^\ominus$ value for the lefthand electrode and add the revised values to get the cell e.m.f. and the polarity of the lefthand electrode.

As an example, we will calculate the e.m.f. of the cell which has electrode systems

$$Pb(s)\,|\,Pb^{2+}(aq) \text{ and } Mg(s)\,|\,Mg^{2+}(aq)$$

The cell diagram can be written

$$Mg(s)\,|\,Mg^{2+}(aq)\,\vdots\,Pb^{2+}(aq)\,|\,Pb(s)$$

From tables

$$Mg^{2+}(aq) \mid Mg(s);\ E^{\ominus} = -2.37\ V$$
$$Pb^{2+}(aq) \mid Pb(s);\quad E^{\ominus} = -0.13\ V$$

Reverse the sign of $E^{\ominus}$ value for lefthand electrode and add

$$+2.37\ V - 0.13\ V = +2.24\ V$$

The e.m.f of the cell, under standard concentration conditions, will be 2.24 V at 25 °C and the lead electrode will be the positive pole. If other than standard ion concentrations are used, the E values for the separate half-cells must be calculated using the Nernst equation.

2 *Predicting whether a reaction is likely to take place*.– This will be discussed in Topic 16 (d-block elements).

15.4 **Acid-base equilibria**

Reactions in which acids and bases are involved have been discussed in Topic 6, where the Lowry-Brønsted theory of acid-base behaviour was introduced. In this theory an *acid* is defined as a substance which can *provide protons* under reaction conditions, and a *base* as a substance which can *combine with protons*.

The chemistry of acid-base systems is concerned with equilibria between electrovalently-bonded species and covalently-bonded species. Equal sharing of electrons (a true covalent bond) occurs only between like atoms, as in H_2, Cl_2, etc. Bonding between unlike atoms always results in unequal sharing and polar bonds, e.g. $\overset{\delta+}{H}—\overset{\delta-}{Cl}$.

When sharing is carried to its limit, transfer of an electron pair takes place and an electrovalent bond results. The degree of electron sharing in polar molecules is changed when they are dissolved in polar solvents, often resulting in the formation of an electrovalent bond. This happens when hydrogen chloride reacts with water

$$HCl(g) + H_2O(l) \rightarrow H_3O^+ + Cl^-(aq),$$

a process which can be resolved into two stages.

$$HCl(g) \rightarrow H^+ + Cl^-$$
$$H^+ + Cl^- + H_2O(l) \rightarrow H_3O^+ + Cl^-(aq)$$

The hydrogen ion (H^+) is a single proton, with no electrons associated with it to produce a comparatively large volume, hence it is some 50 000 times smaller than the next smallest atom. The possibility of very close approach between the free proton and the oxygen atom in the water molecule results in a strong bond being formed by the lone pair of electrons on the oxygen atom. Many other substances react with water in this way.

In acid-base systems we have a competition for protons in a precisely similar way to the competition for electrons in redox systems. Thus the equilibria involved in the reaction of hydrogen chloride with water:

$$\underset{(\text{acid}_1)}{HCl(g)} \rightleftharpoons \underset{(\text{base}_1)}{Cl^-(aq)} + H^+$$

and

$$\underset{(\text{acid}_2)}{H_3O^+} \rightleftharpoons \underset{(\text{base}_2)}{H_2O(l)} + H^+$$

are similar to

$$\underset{(\text{reductant}_1)}{Zn(s)} \rightleftharpoons \underset{(\text{oxidant}_1)}{Zn^{2+}(aq)} + 2e^-$$

and

$$\underset{(\text{reductant}_2)}{Cu(s)} \rightleftharpoons \underset{(\text{oxidant}_2)}{Cu^{2+}(aq)} + 2e^-$$

In redox systems the relative tendencies of the two reductants to form ions in solution determines which of them has the greater reducing strength. In the case above zinc has the greater reducing strength and the equilibrium

$$\underset{(\text{reductant}_1)}{Zn(s)} + \underset{(\text{oxidant}_2)}{Cu^{2+}(aq)} \rightleftharpoons \underset{(\text{oxidant}_1)}{Zn^{2+}(aq)} + \underset{(\text{reductant}_2)}{Cu(s)}$$

is well over to the right. A similar state of affairs is found in the acid-base equilibrium

$$\underset{(\text{acid}_1)}{HCl(g)} + \underset{(\text{base}_2)}{H_2O(l)} \rightleftharpoons \underset{(\text{base}_1)}{Cl^-(aq)} + \underset{(\text{acid}_2)}{H_3O^+}$$

but here the position of equilibrium is determined by the relative tendencies of the acids and bases to lose and gain protons. In this particular equilibrium the concentrations of $Cl^-(aq)$ and $H_3O^+(aq)$ greatly exceed those of $HCl(g)$. This means that hydrogen chloride loses protons much more readily than H_3O^+, and that $H_2O(l)$ accepts protons more readily than $Cl^-(aq)$.

Towards soluble bases water can act as an acid. In aqueous ammonia, for example, the acid-base equilibrium is

$$\underset{(\text{base}_1)}{NH_3(aq)} + \underset{(\text{acid}_2)}{H_2O(l)} \rightleftharpoons \underset{(\text{acid}_1)}{NH_4^+(aq)} + \underset{(\text{base}_2)}{OH^-(aq)}$$

In this solution there is a higher concentration of $NH_3(aq)$ than of $NH_4^+(aq)$, indicating that $NH_4^+(aq)$ loses protons more readily than $H_2O(l)$, and that

OH^-(aq) accepts protons more readily than NH_3 (aq). The competing equilibria are

$$NH_4^+(aq) \rightleftharpoons NH_3(aq) + H^+$$
and $$H_2O(l) \rightleftharpoons OH^-(aq) + H^+$$

If an aqueous solution of hydrogen chloride is added to an aqueous solution of ammonia, the acid H_3O^+ reacts with the base OH^-(aq) to set up the equilibrium

$$H_3O^+ + OH^-(aq) \rightleftharpoons 2H_2O\,(l)$$

This is always well over to the righthand side and leads to removal of H_3O^+(aq) and OH^-(aq) ions from solution, thus altering the equilibrium position of

$$HCl(g) + H_2O(l) \rightleftharpoons H_3O^+ + Cl^-(aq)$$
and $$NH_3(aq) + H_2O(l) \rightleftharpoons NH_4^+(aq) + OH^-(aq)$$

towards the righthand side also. The final products are mainly ammonium ions and chloride ions in aqueous solution.

Similar equilibria are set up in other solvents. In liquid ammonia, for example, there is the equilibrium

$$2NH_3(l) \rightleftharpoons NH_4^+ + NH_2^-$$

corresponding to $2H_2O(l) \rightleftharpoons H_3O^+ + OH^-$ in water. Reactions in liquid ammonia are similar to those in water. For example, the compound sodamide, $NaNH_2$, ionizes

$$NaNH_2 \rightarrow Na^+ + NH_2^-$$

and ammonium chloride also ionizes

$$NH_4Cl \rightarrow NH_4^+ + Cl^-$$

A 'neutralization' reaction then takes place

$$NH_4^+ + NH_2^- \rightarrow 2NH_3(l).$$

In this section we shall deal with aqueous systems only, but the wider application of the principles involved should not be forgotten.

The strengths of acids and bases

In order to compare the relative tendencies of metals to form ions in solution, and thus to compare their strengths as reductants, a standard electrode system (the hydrogen electrode) is used. The principle underlying this method is to put the equilibrium between a metal and its ions

$$M(s) \rightleftharpoons M^{z+}(aq) + ze^-$$

in competition with the equilibrium

$$2H^+(aq) + 2e^- \rightleftharpoons H_2(g)$$

The effect of the competing equilibria on each other is then measured by finding the e.m.f. of a voltaic cell.

Can a similar method of comparing the relative strength of ions and molecules as acids and bases be found? In general, acidic and basic properties are important mainly in connection with solutions, the commonest solvent being water. If a substance (represented by HA) which can function as an acid is dissolved in water, two competing equilibria are established:

$$HA(aq) \rightleftharpoons A^-(aq) + H^+$$

and $$H^+ + H_2O(l) \rightleftharpoons H_3O^+.$$

If HA loses protons readily, a high concentration of H_3O^+ ions will be produced when the system has reached equilibrium. In this case HA is functioning as a strong acid. On the other hand, if HA is a weak acid, with no pronounced tendency to part with protons, the concentration of H_3O^+ ions will be smaller. In effect, the H_3O^+ ion can be used as a standard against which to compare the relative strengths of acids. The problem now becomes one of measuring $[H_3O^+]_{eqm}$ in aqueous solutions of different acids.

(There is no general agreement about the name to be used for the H_3O^+ ion; it has been called the hydronium ion, the hydroxonium ion, and the oxonium ion. The first name, the hydronium ion, is that recommended by IUPAC. Unless special attention needs to be directed towards the hydronium ion as a combination of a hydrogen ion and a water molecule the usual practice is to use the symbol $H^+(aq)$ and to call this 'the hydrogen ion'.)

One way of comparing hydrogen ion concentrations is to measure pH values, as has been done earlier in the course using indicators. The indicator method gives rough values only; we now need a more accurate method.

The relationship between the pH value of a solution and the hydrogen ion concentration is

$$pH = -\lg [H^+(aq)]$$

where $[H^+(aq)]$ is measured in mol dm^{-3}. A logarithmic scale is used because the range of possible hydrogen ion concentrations in solution is very large (from about 10 to 10^{-15}). The minus sign is introduced to make pH values positive in almost all cases encountered in practice. A few examples will help to make the relationship clearer.

When $[H^+(aq)] = 10^{-3}$ mol dm^{-3}; pH = 3 ($-\lg [H^+(aq)] = +3$)
$[H^+(aq)] = 10^{-8}$ mol dm^{-3}; pH = 8
$[H^+(aq)] = 5 \times 10^{-6}$ mol dm^{-3}; pH = 5.3

The last example may surprise you; remember that the mantissa of a logarithm (the figures after the decimal point) is *always* positive, whereas the characteristic (before the decimal point) can be positive or negative.

$$\lg 5 \times 10^{-6} = \bar{6}.7 = -6 + .7 = -5.3, \text{ therefore pH} = 5.3.$$

Measurement of $[H^+(aq)]$ by electrical methods

In principle, the simplest method for measuring hydrogen ion concentration is to use a hydrogen electrode in the solution of unknown $H^+(aq)$ concentration as a half-cell. This can then be combined with a standard electrode to form a complete cell, the e.m.f. of which can be measured. The value of $[H^+(aq)]$ can then be calculated, using the Nernst equation.

For example, if the cell

$$Pt[H_2(g)] \mid \underset{\text{(Concentration unknown)}}{2H^+(aq)} \,\vdots\, Cu^{2+}(aq, M) \mid Cu(s)$$

is set up and the e.m.f. found to be 0.43 V. Then
for the copper electrode, $E = E^{\ominus} = 0.34$ V (from tables)
$\therefore$ for the hydrogen electrode $E = 0.34 - 0.43 = -0.09$ V

but, as $E^{\ominus} = 0$, $E = 0.06 \lg [H^+(aq)]$

$$\therefore \lg [H^+(aq)] = -\frac{0.09}{0.06} = -1.5$$

$\therefore$ pH $= -\lg [H^+(aq)] = -(-1.5) = 1.5$
also $[H^+(aq)] = \text{antilg}\,(-1.5) = \text{antilg}\,\bar{2}.5$
$= 3 \times 10^{-2}$ mol dm^{-3}

As you will have seen, however, the hydrogen electrode is not easy to use. It is bulky when the hydrogen generator is taken into account, slow to reach equilibrium, and rather easily 'poisoned' by impurities. Alternative electrodes have therefore been sought, and, up to the present, the *glass electrode* is the most successful of these.

The glass electrode consists of a thin-walled bulb blown from special glass of low melting point. A solution of constant pH (a 'buffer' solution of which details will be given later) is placed inside the bulb with a platinum wire dipping into it. If the bulb is now immersed in a solution of unknown pH a potential is developed on the platinum wire and the whole arrangement can be used as a half-cell. Combination with a suitable reference electrode enables e.m.f measurements to be made. The resistance of the glass bulb is high (10^7–10^8 ohm) and a very sensitive voltmeter must be used to measure the e.m.f. In practice a valve (or transistor) voltmeter is used. The reference electrode is usually either a *calomel electrode* (a platinum wire dipping into mercury below a solution of mercury(I) chloride in saturated potassium chloride solution), or a *silver/silver chloride electrode* (a silver wire coated with silver chloride dipping into saturated potassium chloride solution).

The theory of the glass electrode is complicated but an arrangement such as

Pt | solution (A) of known pH | glass bulb | solution (B) of unknown pH ⁞ reference electrode

can be attached to a valve voltmeter and calibrated by using solutions of known pH in place of solution (B). An instrument designed on this basis is called a *pH meter*. In commercial pH meters the glass electrode and the reference electrode may be combined in one unit which can be dipped into the solution under investigation. The voltmeter scale is calibrated to read directly in pH units.

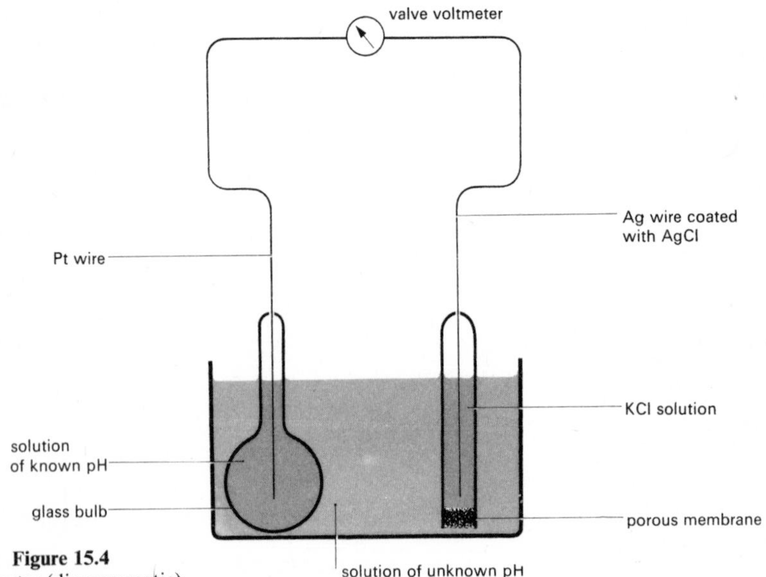

Figure 15.4
A pH meter (diagrammatic).

pH values in aqueous solutions

Even in alkaline solutions of high [OH^-(aq)] it is possible to detect and measure a hydrogen ion concentration. This is due to the equilibrium

$$H_2O(l) \rightleftharpoons H^+(aq) + OH^-(aq)$$

existing in all aqueous solutions. For this equilibrium, as shown earlier

$$K_w = [H^+(aq)]_{eqm}[OH^-(aq)]_{eqm}$$

At 25 °C, $K_w = 10^{-14}$ mol² dm⁻⁶, so that when

$$[H^+(aq)]_{eqm} = [OH^-(aq)]_{eqm}$$

the value of each is $\sqrt{10^{-14}} = 10^{-7}$ mol dm^{-3}. This equality occurs in pure water (which is very difficult to obtain) and in any absolutely neutral solution; for these systems

$$pH = -\lg 10^{-7} = 7$$

A solution of pH < 7 is acidic; $[H^+(aq)]_{eqm} > [OH^-(aq)]_{eqm}$

A solution of pH > 7 is basic (or alkaline); $[H^+(aq)]_{eqm} < [OH^-(aq)]_{eqm}$

For example, in 0.1M HCl,

$$[H^+(aq)]_{eqm} = 10^{-1} \text{ mol dm}^{-3}$$
$$\therefore pH = 1$$
and $$[OH^-(aq)]_{eqm} = 10^{-13} \text{ mol dm}^{-3}$$

In 0.01M NaOH (assumed completely dissociated into ions)

$$[OH^-(aq)]_{eqm} = 10^{-2} \text{ mol dm}^{-3}$$
$$\therefore [H^+(aq)]_{eqm} = 10^{-12} \text{ mol dm}^{-3}$$
$$\therefore pH = 12$$

Theoretically, assuming complete ionization, pure sulphuric acid (which is 18M) should have a pH of about −1.5. In practice, however, most of the acid at this concentration is in the form of covalent molecules, ionization is low, and the pH value is correspondingly high. The range of pH values actually encountered in aqueous systems varies between just less than zero (negative value) and a little over 14 (concentrated alkaline solutions).

Change of pH with dilution and its relationship to the strength of an acid

If we use a pH meter to measure pH values for progressively diluted hydrochloric acid, we find that there is an increase of about one pH unit per ten-fold dilution, e.g.

pH of M HCl(aq) is approximately 0
pH of 0.1M HCl(aq) is approximately 1
pH of 0.01M HCl(aq) is approximately 2
pH of 0.001M HCl(aq) is approximately 3

These observations can be accounted for if we assume that the acid is almost completely ionized at all concentrations, so that the equilibrium

$$HCl(aq) + H_2O(l) \rightleftharpoons H_3O^+ + Cl^-(aq)$$

lies almost completely over to the right. A ten-fold dilution will then reduce the value of $[H^+(aq)]_{eqm}$ by 10 and the pH will increase by lg 10 = 1. A few other acid solutions behave in this way but for most acidic solutions the increase in pH for a dilution factor of ten is less than one unit, and the pH values for comparable concentrations are always greater than for hydrochloric acid solutions. This could mean that dissociation into ions is incomplete and a considerable proportion of reactants remains when the equilibrium

$$HA(aq) + H_2O(l) \rightleftharpoons H_3O^+ + A^-(aq)$$

is attained. (HA is used to stand for any acid which behaves in this way.) Thus the hydrogen ion concentration will be smaller than would be expected for complete dissociation and the pH value higher. Further dilution (increase of $H_2O(l)$ concentration) would then move the equilibrium more towards the right until, at high dilution, conversion to products (ions) is virtually complete.

Acids which ionize nearly completely at moderate dilutions (0.1M or 0.01M) are called *strong acids*. Those which ionize slightly, or exist mainly as the covalently bonded form under these conditions, are called *weak acids*. As with electrovalent and covalent bonds, there is no sharp dividing line between strong and weak acids but rather a kind of spectrum of acidic properties.

The arguments outlined above can be pursued quantitatively. We shall consider a relatively weak acid and write the general equilibrium equation as

$$HA(aq) \rightleftharpoons H^+(aq) + A^-(aq)$$

so that we can neglect the $[H_2O(l)]_{eqm}$ term.

For this equilibrium

$$K_c = \frac{[H^+(aq)]_{eqm}[A^-(aq)]_{eqm}}{[HA(aq)]_{eqm}}$$

When dealing with acid-base equilibria the symbol K_a is often used instead of K_c

$$\frac{[H^+(aq)]_{eqm}[A^-(aq)]_{eqm}}{[HA(aq)]_{eqm}} = K_a \text{ (the \textit{dissociation constant} of the acid)}$$

Using this expression we can calculate the value of K_a for an acid solution if we know the concentration of the solution and its pH. A specific example will show the method.

The pH of a 0.01M solution of formic acid (HCO_2H) is 2.90 at 25 °C. Calculate the value of K_a at this concentration and temperature.

$$HCO_2H(aq) \rightleftharpoons HCO_2^-(aq) + H^+(aq)$$

$$K_a = \frac{[HCO_2^-(aq)]_{eqm}[H^+(aq)]_{eqm}}{[HCO_2H(aq)]_{eqm}}$$

Neglecting the hydrogen ions which arise from ionization of the water, since the concentration of these will be very small compared with the concentration of those from the acid, we can say that

$$[H^+(aq)]_{eqm} = [HCO_2^-(aq)]_{eqm}$$

and $$[HCO_2H(aq)]_{eqm} = 0.01 - [H^+(aq)] \text{ mol dm}^{-3}$$

$$pH = -\lg [H^+(aq)]_{eqm} = 2.90$$

$$\therefore \lg [H^+(aq)]_{eqm} = -2.90$$

and $$[H^+(aq)]_{eqm} = \text{antilg}(-2.90) = \text{antilg } \bar{3}.10$$
$$= 1.26 \times 10^{-3} \text{ mol dm}^{-3}$$
$$= [HCO_2^-(aq)]_{eqm}$$

and $$[HCO_2H(aq)]_{eqm} = 0.01 - 1.26 \times 10^{-3}$$
$$= 10 \times 10^{-3} - 1.26 \times 10^{-3}$$
$$= 8.74 \times 10^{-3} \text{ mol dm}^{-3}$$

$$K_a = \frac{(1.26 \times 10^{-3})^2}{8.74 \times 10^{-3}} = 1.81 \times 10^{-4} \text{ mol dm}^{-3}$$

If we measure the pH of a weak acid at different concentrations, calculate values of K_a from these, and find K_a to be fairly constant, we would see that our assumptions about the increasing dissociation of a weak acid on dilution are not unreasonable.

By reversing the calculation, the pH of a solution of a weak acid for any molarity can be found if the value of K_a for the acid is known. Can you do this for the following example?

Calculate the pH of a 0.001M solution of aminoacetic acid, $NH_2CH_2CO_2H$ ($K_a = 1.7 \times 10^{-10}$ mol dm^{-3}).

15.5 Buffer solutions and indicators

Buffer solutions

Acetic acid is a weak acid, being only slightly ionized in solution ($K_a = 2 \times 10^{-5}$ mol dm^{-3}). In an aqueous solution of acetic acid the equilibrium

$$\underset{\text{(acid)}}{CH_3CO_2H(aq)} \rightleftharpoons \underset{\text{(corresponding base)}}{CH_3CO_2^-(aq)} + H^+(aq) \qquad (1)$$

lies well over to the left, and the hydrogen ion concentration is relatively small. There is a second equilibrium involving hydrogen ions in this system

$$H^+(aq) + OH^-(aq) \rightleftharpoons H_2O(l) \qquad (2)$$

What happens when additional acetate ions are added to the solution? This could be done by adding a soluble salt of acetic acid, such as sodium acetate, which is highly ionized in solution. Using Le Chatelier's principle, equilibrium (1) moves to the left and the hydrogen ion concentration decreases. This we should expect anyway since we are adding a base (the acetate ion) to an acid. The mixture now contains a relatively high concentration of un-ionized acetic acid and a relatively high concentration of acetate ion. It therefore contains both an acid and a base.

If more hydrogen ions are added to this system, by adding a small volume of a solution of a strong acid, these will combine with acetate ions to form more un-ionized acetic acid. Equilibrium (1) moves to the left, removing nearly all the added hydrogen ions. The concentration of hydrogen ions, and thus the pH of the solution, will alter a little, but not very much.

Addition of a strong base, for example sodium hydroxide, to the system disturbs equilibrium (2) so that $OH^-(aq)$ ions combine with $H^+(aq)$ ions to form $H_2O(l)$. This reduces the hydrogen ion concentration, and more CH_3CO_2H molecules ionize to restore it to near its original value. The two equilibria adjust themselves in this way until nearly all the added hydroxide ions are removed. The pH value of the system will rise a little in consequence, but not very much. The changes in pH resulting from additions of acid or base are much smaller than they would be if the mixture of weak acid and salt was not present.

Solutions of this kind, containing a weak acid and its corresponding base (sometimes called the conjugate base), thus provide a 'buffer' against the effects of adding strong acid or strong base. They are therefore known as *buffer solutions*. Essentially they are solutions possessing readily available reserve supplies of both an acid and a base.

Another example of a buffer solution contains a mixture of ammonium chloride (highly ionized) and ammonia (present mainly as neutral molecules). The equilibria present are

$$NH_4^+(aq) \rightleftharpoons NH_3(aq) + H^+(aq)$$
(acid) (corresponding base)

and $H^+(aq) + OH^-(aq) \rightleftharpoons H_2O(l)$

Addition of more $H^+(aq)$ ions results in their reaction with the base $NH_3(aq)$ to form $NH_4^+(aq)$ ions. When more $OH^-(aq)$ ions are added the following changes occur

$$H^+(aq) + OH^-(aq) \rightarrow H_2O(l)$$
$$NH_4^+(aq) \rightarrow NH_3(aq) + H^+(aq)$$

until the two equilibria are again restored. Again, the pH value remains nearly constant.

Calculations involving buffer solutions

If the relative concentrations of acid and base in a buffer solution are known, the pH of the mixture can be calculated. Alternatively, the composition of the mixture needed to make a buffer solution of a given pH value can be found.

An equation that enables these calculations to be done can be derived quite easily. We will consider the equilibrium

$$CH_3CO_2H(aq) \rightleftharpoons CH_3CO_2^-(aq) + H^+(aq)$$

for which

$$\frac{[CH_3CO_2^-(aq)]_{eqm}[H^+(aq)]_{eqm}}{[CH_3CO_2H(aq)]_{eqm}} = K_a$$

re-arranging this, we have

$$[H^+(aq)]_{eqm} = K_a\left(\frac{[CH_3CO_2H(aq)]_{eqm}}{[CH_3CO_2^-(aq)]_{eqm}}\right)$$

taking logarithms of both sides

$$\lg\,[H^+(aq)]_{eqm} = \lg K_a + \lg\left(\frac{[CH_3CO_2H(aq)]_{eqm}}{[CH_3CO_2^-(aq)]_{eqm}}\right)$$

change the signs all through

$$-\lg\,[H^+(aq)]_{eqm} = -\lg K_a - \lg\left(\frac{[CH_3CO_2H(aq)]_{eqm}}{[CH_3CO_2^-(aq)]_{eqm}}\right)$$

now $$-\lg [H^+(aq)]_{eqm} = pH$$

$$\therefore pH = -\lg K_a - \lg\left(\frac{[CH_3CO_2H(aq)]_{eqm}}{[CH_3CO_2^-(aq)]_{eqm}}\right)$$

A similar argument applied to the equilibrium

$$NH_4^+(aq) \rightleftharpoons NH_3(aq) + H^+(aq)$$

leads to the equation

$$pH = -\lg K_a - \lg\left(\frac{[NH_4^+(aq)]_{eqm}}{[NH_3(aq)]_{eqm}}\right)$$

The general equation is

$$pH = -\lg K_a - \lg\left(\frac{[acid]_{eqm}}{[base]_{eqm}}\right)$$

Two points are worth noting from this equation.

1 The pH of a buffer solution depends on the *ratio* of the concentrations of acid and base, not on the actual values of these concentrations.

2 When $[acid]_{eqm} = [base]_{eqm}$, $pH = -\lg K_a$. This means that the K_a value of an acid can be found by measuring the pH value of a solution of the acid which has been half neutralized by a strong base (experiment 15.5).

Two examples will show how the equation can be used.

Example 1

Calculate the pH value of a solution which is 0.05 molar with respect to acetic acid and 0.20 molar with respect to sodium acetate (K_a for acetic acid $= 2 \times 10^{-5}$ mol dm^{-3}).

In this, and all similar calculations, we are dealing with systems for which K_a is small. We can therefore simplify the calculations by assuming that the acetic acid is un-ionized and that all the acetate ions come from the sodium acetate.

By doing this we can write

$$[acid]_{eqm} = [CH_3CO_2H(aq)]_{eqm} = 0.05 \text{ mol dm}^{-3}$$

and $$[base]_{eqm} = [CH_3CO_2^-(aq)]_{eqm} = 0.20 \text{ mol dm}^{-3}$$

$$\therefore \text{pH} = -\lg 2 \times 10^{-5} - \lg \frac{0.05}{0.20}$$

$$= 4.7 - \lg \frac{0.05}{0.20}$$
$$= 4.7 - \bar{1}.4$$
$$= 4.7 + 0.6$$
$$= 5.3.$$

Example 2

In what proportions must 0.1M solutions of ammonia and ammonium chloride be mixed to obtain a buffer solution of pH 9.8 (K_a for the ammonium ion is 6×10^{-10} mol dm^{-3}). Here the acid is $NH_4^+(aq)$ and the base NH_3.

$$\therefore 9.8 = -\lg 6 \times 10^{-10} - \lg \left(\frac{[NH_4^+(aq)]_{eqm}}{[NH_3(aq)]_{eqm}} \right)$$

$$9.8 = 9.2 - \lg \left(\frac{[NH_4^+(aq)]_{eqm}}{[NH_3(aq)]_{eqm}} \right)$$

$$\therefore 0.6 = -\lg \left(\frac{[NH_4^+(aq)]_{eqm}}{[NH_3(aq)]_{eqm}} \right)$$

$$\text{or } \lg \left(\frac{[NH_4^+(aq)]_{eqm}}{[NH_3(aq)]_{eqm}} \right) = -0.6 = \bar{1}.4$$

$$\therefore \frac{[NH_4^+(aq)]_{eqm}}{[NH_3(aq)]_{eqm}} = 0.25$$

The solutions must therefore be mixed in the proportions, 0.25 volume 0.1M ammonium chloride to 1 volume 0.1M ammonia solution.

The effect of adding $H^+(aq)$ or $OH^-(aq)$ ions to a buffer solution can be seen by considering what will happen when these are added to a buffer and a non-buffer solution. For the buffer solution we can take a solution that is 0.1M with respect to both ammonium chloride and ammonia. For this $[NH_4^+(aq)]_{eqm} = [NH_3(aq)]_{eqm}$ and pH $= -\lg K_a = 9.2$. For the non-buffer solution we will take a 10^{-5}M solution of sodium hydroxide, which has pH $= 9$.

What happens if we add 10 cm^3 of a molar solution of a strong acid, $[H^+(aq)] = 1$, to one cubic decimetre of each solution? The actual amount of $H^+(aq)$ that is being added $= \frac{10}{1000} = 0.01$ mol. If the *buffer solution* works perfectly, all this additional hydrogen ion will be removed by the reaction

$$NH_3(aq) + H^+(aq) \rightarrow NH_4^+(aq)$$

At the same time 0.01 mol $NH_3(aq)$ will be removed and 0.01 mol $NH_4^+(aq)$ formed. We now have

$$[NH_4^+(aq)]_{eqm} = 0.1+0.01 = 0.11 \text{ mol dm}^{-3}$$

and $$[NH_3(aq)]_{eqm} = 0.1-0.01 = 0.09 \text{ mol dm}^{-3}$$

and $$\begin{aligned} pH &= 9.2 - \lg\frac{0.11}{0.09} \\ &= 9.2-0.09 \\ &= 9.11, \end{aligned}$$ *a change of less than* 0.1 *pH unit*

For the non-buffer solution, the value of $[H^+(aq)]_{eqm} = 10^{-9}$ mol dm^{-3}, before the acid is added, and becomes $10^{-9}+0.01$ mol dm^{-3} after the acid is added. Clearly the new value of $[H^+(aq)]$ is very nearly 0.01 or 10^{-2} mol dm^{-3}. For the new solution pH $= -\lg 10^{-2} = 2$. There is therefore *a change of 7 pH units.*

Now see if you can work out the pH change resulting from the addition of 10 cm^3 of M sodium hydroxide solution, $[OH^-(aq)] = 1$ mol dm^{-3}, to the buffer and non-buffer solutions.

Many chemical processes, especially those that take place in living organisms, are sensitive to pH changes and either cease or proceed in undesired directions if the pH of the system changes markedly. In such systems buffer solutions provide a means of establishing conditions of fairly constant pH value. Some acid-base chemistry in the human body in which buffer solutions play an important part is discussed in the background reading which can be found at the end of this topic.

Indicators

You will have used acid-base indicators such as methyl orange, methyl red, phenolphthalein, and litmus, in previous work. They are used to test for alkalinity and acidity, and for detecting the end-point in acid-base titrations. A given indicator cannot be used in all circumstances; some are more suitable for use with weak acids and others with weak bases. From a study of titration curves (see appendix II) it is apparent that the rate of change of pH value is related to the rate of indicator colour change. Phenolphthalein, for example, is unsuitable for use when titrating an acid with a weak base, as in the reaction between HCl(aq) and $NH_3(aq)$, because the colour change takes place over the range pH 8–10, and it is in this area that the pH of this system changes very slowly. Methyl red is suitable for this titration since the colour change with this indicator (red → yellow) takes place in the pH range 4–7 and it is here that the rate of pH change is most rapid. This makes it easy to determine the end-point.

An indicator may be considered as a weak acid, for which either the acid or the corresponding base, or both, are coloured. We can represent this in a general way, using HIn for the acid form, as

$$\underset{\text{colour A}}{HIn(aq)} \rightleftharpoons H^+(aq) + \underset{\text{colour B}}{In^-(aq)}$$

Addition of acid, $H^+(aq)$ ions for example, displaces the equilibrium to the left and increases the intensity of colour A. Addition of base, $OH^-(aq)$ or $NH_3(aq)$, removes hydrogen ions,

$$H^+(aq) + OH^-(aq) \rightarrow H_2O(l)$$

or $$NH_3(aq) + H^+(aq) \rightarrow NH_4^+(aq),$$

with the result that the equilibrium moves to the right to restore the value of K_a, and increase the intensity of colour B. The colour of the system will depend on the relative concentrations, $[HIn(aq)]_{eqm}$ and $[In^-(aq)]_{eqm}$.

Regarding indicators as weak acids we can apply the general equation

$$pH = -\lg K_a - \lg \frac{[\text{acid}]}{[\text{base}]}$$

to them. That is, for an indicator HIn,

$$pH = -\lg K_a - \lg \frac{[HIn(aq)]_{eqm}}{[In^-(aq)]_{eqm}}$$

If the ratio of $[HIn(aq)]$ to $[In^-(aq)]$ at a given point in the colour range of an indicator is known, together with the pH value for this mixture, the value of K_a for an indicator can be calculated (see experiment 15.5).

Taking phenolphthalein as an example, we can explore the possible pH range over which the colour change can be perceived. For this indicator K_a is approximately 10^{-9}. Representing phenolphthalein by HPh in the acid form, the equilibrium is

$$\underset{\text{colourless}}{HPh(aq)} \rightleftharpoons H^+(aq) + \underset{\text{red}}{Ph^-(aq)}$$

When the indicator has developed half the full red colour,

$$[HPh(aq)]_{eqm} = [Ph^-(aq)]_{eqm}$$

and $$pH = -\lg K_a = -\lg 10^{-9} = 9$$

In a solution (or very pure water) of pH 7, the ratio of red (base) to colourless (acid) form can be calculated

$$7 = -\lg K_a - \lg \frac{[\mathrm{HPh(aq)}]_{\mathrm{eqm}}}{[\mathrm{Ph^-(aq)}]_{\mathrm{eqm}}}$$

$$\therefore 7 = 9 - \lg \frac{[\mathrm{HPh(aq)}]_{\mathrm{eqm}}}{[\mathrm{Ph^-(aq)}]_{\mathrm{eqm}}}$$

$$\therefore \lg \frac{[\mathrm{HPh(aq)}]_{\mathrm{eqm}}}{[\mathrm{Ph^-(aq)}]_{\mathrm{eqm}}} = 2$$

and $$\frac{[\mathrm{HPh(aq)}]_{\mathrm{eqm}}}{[\mathrm{Ph^-(aq)}]_{\mathrm{eqm}}} = \frac{100}{1}$$

Out of 101 molecules, 100 will be in the colourless form and one only in the red form. The eye cannot detect this small proportion of colour so the mixture appears colourless.

At pH 8

$$\lg \frac{[\mathrm{HPh(aq)}]_{\mathrm{eqm}}}{[\mathrm{Ph^-(aq)}]_{\mathrm{eqm}}} = 9 - 8 = 1$$

and $$\frac{[\mathrm{HPh(aq)}]_{\mathrm{eqm}}}{[\mathrm{Ph^-(aq)}]_{\mathrm{eqm}}} = \frac{10}{1}$$

This ratio of colourless to coloured form is just detectable.

At pH 10

$$\lg \frac{[\mathrm{HPh(aq)}]_{\mathrm{eqm}}}{[\mathrm{Ph^-(aq)}]_{\mathrm{eqm}}} = 9 - 10 = -1$$

and $$\frac{[\mathrm{HPh(aq)}]_{\mathrm{eqm}}}{[\mathrm{Ph^-(aq)}]_{\mathrm{eqm}}} = \frac{1}{10}$$

Out of 11 molecules of indicator, 10 will be in the coloured form. Any increase in the proportion of coloured molecules will not be noticed as an increase in colour.

Most indicators have a detectable colour change over a range of about 2 pH units. This is a consequence of the sensitivity of our eyes to colour changes.

Experiment 15.5

To measure K_a for (*a*) an indicator and (*b*) a weak acid

For an indicator, HIn

$$pH = -\lg K_a - \lg \frac{[HIn]_{eqm}}{[In^-]_{eqm}}$$

If the ratio [HIn]/$[In^-]$ is found for a solution of known pH, K_a can be calculated.

Similarly, for a weak acid, HA

$$pH = -\lg K_a - \lg \frac{[HA]_{eqm}}{[A^-]_{eqm}}$$

In a buffer solution containing equimolar concentrations of HA and A^-,

$$pH = -\lg K_a,$$

and K_a can be calculated if the pH is known.

In this experiment K_a for bromophenol blue and K_a for benzoic acid are determined.

Procedure

Make up the following solutions:
Solution X Add one drop of concentrated hydrochloric acid to 5 cm^3 of bromophenol blue solution.
Solution Y Add one drop of 4M sodium hydroxide solution to 5 cm^3 of bromophenol blue solution.
(Solution X contains the indicator wholly in the HIn form and solution Y has the same concentration of indicator, wholly in the In^- form.)

a. To measure K_a *for bromophenol blue* – Arrange 18 test-tubes in nine pairs, one behind the other, in a double test-tube rack so that, when looking through a pair of tubes the colour seen will be that due to the solutions in both tubes.

Put 10 cm^3 distilled water into each of the 18 test-tubes and add drops of solutions X and Y, as follows:

Tube	1	2	3	4	5	6	7	8	9
Drops of X	1	2	3	4	5	6	7	8	9

Tube	10	11	12	13	14	15	16	17	18
Drops of Y	9	8	7	6	5	4	3	2	1

Add 5 cm^3 0.02M formic acid to 5 cm^3 0.02M sodium formate in a test-tube of the same size as those in the rack (K_a for formic acid is 2×10^{-4} mol dm^{-3}). Add 10 drops of bromophenol blue solution to the test-tube and shake to mix the contents of the tube. Compare the colour of the mixture with the colours seen by looking through the *pairs* of the test-tubes in the rack.

Questions

1 Which pair of tubes matches most closely the colour in the formic acid/sodium formate mixture?

2 Calculate the pH of the formic acid/sodium formate mixture.

3 What is the ratio $[HIn]_{eqm}/[In^-]_{eqm}$ at this pH?

4 Calculate K_a for bromophenol blue.

b. To measure K_a *for benzoic acid* – Mix 5 cm^3 0.02M benzoic acid and 5 cm^3 0.02M sodium benzoate in a test-tube; add 10 drops bromophenol blue solution. Compare the colour of the mixture with that seen through the pairs of test-tubes in the rack.

Questions

1 Which pair of tubes matches most closely the colour in the benzoic acid/sodium benzoate mixture?

2 What is the ratio $[HIn]_{eqm}/[In^-]_{eqm}$ in the mixture?

3 Calculate the pH of the mixture.

4 Calculate K_a for benzoic acid.

Background reading

Acid-base chemistry in the human body

The chemical processes which occur in the body are complex and often present challenging problems to the biochemist. Some of the mechanisms are reasonably well understood, as is that by which the body regulates the pH value of the blood. This account is only a part of the story.

Although red to the naked eye, blood consists of a pale yellow aqueous fluid, the plasma, in which red and white cells are suspended. The plasma contains proteins, albumen, fibrinogens, and globulins, and also ions such as Na^+, Cl^-, HCO_3^-, and H^+ present in aqueous solution. The red cells contain haemoglobin, an enzyme known as carbonic anhydrase, and ions such as K^+, Cl^-, HCO_3^-, and H^+, also in aqueous solution. The chief functions of the blood are the transport of oxygen and food materials to the body tissues and the carriage of carbon dioxide and other waste products of metabolism from the tissues to the lungs and kidneys, where they are excreted.

Air inhaled into the lungs finds its way into small air sacs, the alveoli, where it is brought into contact with blood across thin membranes. In man, there are some 700 million alveoli each about 2×10^{-4} cm in diameter. The blood arriving in the lungs is known as venous blood and it contains carbon dioxide collected during its passage round the body. Gaseous exchanges occur across the alveolar membranes, oxygen passing into the blood and carbon dioxide leaving it ready for exhalation. The oxygenated or arterial blood is pumped out of the lungs to circulate through the tissues giving up oxygen to them and collecting carbon dioxide. The process is summarized in figure 15.6a.

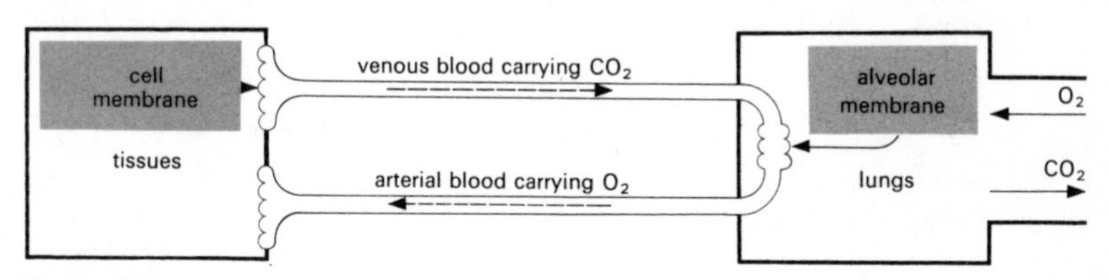

Figure 15.6a
Transport of oxygen and carbon dioxide in the blood.

Let us now examine the chemistry of the events we have been discussing.

Transport of oxygen in the blood

Oxygen is both present in the blood as a 'normal' solution, $O_2(aq)$, and also combines chemically with haemoglobin in the red cells to form a bright red compound, oxyhaemoglobin. Approximately 0.3 per cent of the oxygen is present as $O_2(aq)$ in solution, the rest being in the oxyhaemoglobin. The concentration of the solution of $O_2(aq)$ follows Henry's law.

$$[O_2(aq)] \propto \text{partial pressure of oxygen in alveolar air, } p_{O_2}$$
$$\text{or } [O_2(aq)] = Kp_{O_2}$$

where K is a solubility constant.

Haemoglobin is a complex organic molecule which contains iron, and normally carries a negative charge: we shall represent it here by the symbol Hb^-. The reaction with oxygen is written

$$Hb^-(aq) + O_2(aq) = \underset{\text{oxyhaemoglobin}}{Hb\,O_2^-(aq)}$$

and one mole of oxygen combines with 16 700 g haemoglobin.

The amount of oxygen entering the blood depends upon the partial pressure of oxygen in the alveolar air, which in turn is dependent upon the partial pressure in the atmosphere. In a healthy adult, p_{O_2} in alveolar air is 100 mmHg. The figures in table 15.6 show partial pressures for oxygen at various altitudes and reveal the necessity for breathing equipment in high altitude flight and mountaineering.

Altitude/ft	**Partial pressure of oxygen/mmHg**
sea level	159
10 000	110
20 000	73
50 000	18

Table 15.6
Partial pressure of oxygen in the air

Carriage of carbon dioxide in the blood

When carbon dioxide dissolves in water, it can react slowly to form hydrogen carbonate ions:

$$aq + CO_2(g) \rightarrow CO_2(aq)$$
$$H_2O(l) + CO_2(aq) \rightarrow H^+(aq) + HCO_3^-(aq)$$

The blood carries carbon dioxide in three ways; about 6 per cent is present in solution in the plasma as $CO_2(aq)$, 70 per cent in the form of $HCO_3^-(aq)$, and 24 per cent in the form of a compound which is formed with haemoglobin in the red cells

$$H^+(aq) + Hb^-(aq) + CO_2(aq) \rightarrow HHbCO_2(aq)$$

The normally slow reaction between carbon dioxide and water is catalysed, in both directions, by the enzyme carbonic anhydrase. The efficiency of this catalysis can be appreciated when it is realized that, on average, a red cell spends less than half a second in the alveolar membrane during its passage through the lung.

The formation of HCO_3^- ions in the blood is accompanied by the formation of H^+ ions; hence if the blood were not buffered, its pH would fall when carbon dioxide passes into it. The blood in fact contains a number of buffering systems which in healthy individuals maintains its pH at about 7.4.

Blood buffers

The plasma proteins and haemoglobin can act as buffers because they contain both acidic ($-CO_2H$) and basic ($-NH_2$) groups in their molecules.

The action of a protein molecule as a base and as an acid is represented in figure 15.6b, in the upper diagram as a base and in the lower diagram as an acid.

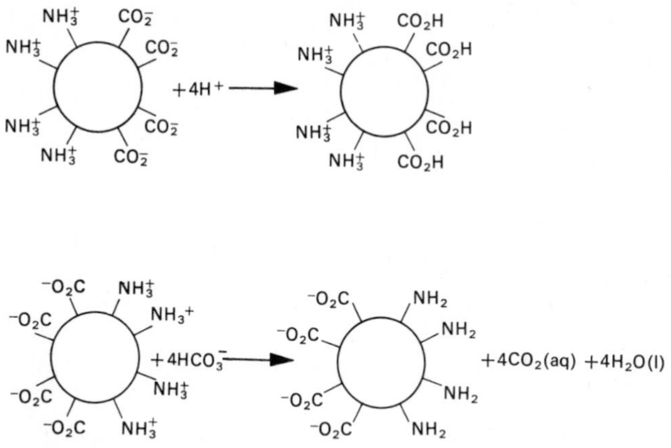

Figure 15.6b

The most important buffer, however, is the hydrogen carbonate ion itself since the equilibrium

$$H_2O(l) + CO_2(aq) \rightleftharpoons H^+(aq) + HCO_3^-(aq)$$

lies well to the left.

The processes which we have been discussing are summarized in figure 15.6c. Notice the phenomenon known as the chloride shift. Can you account for it?

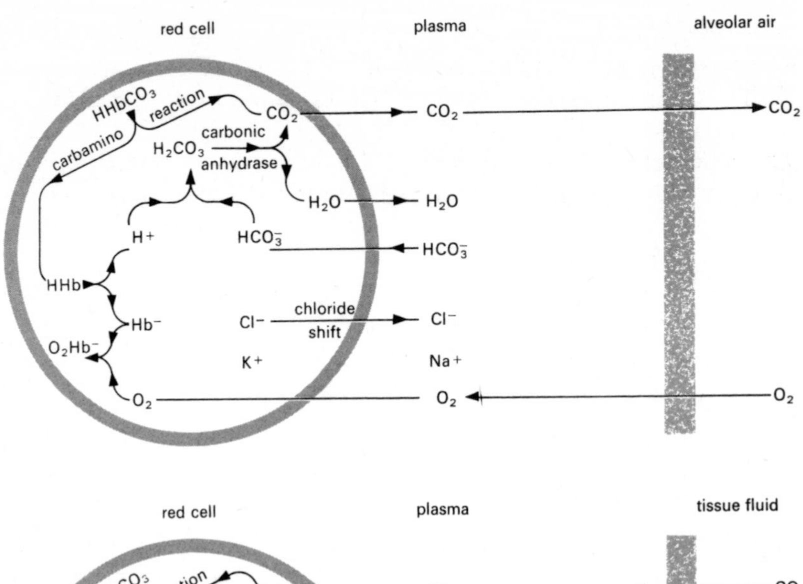

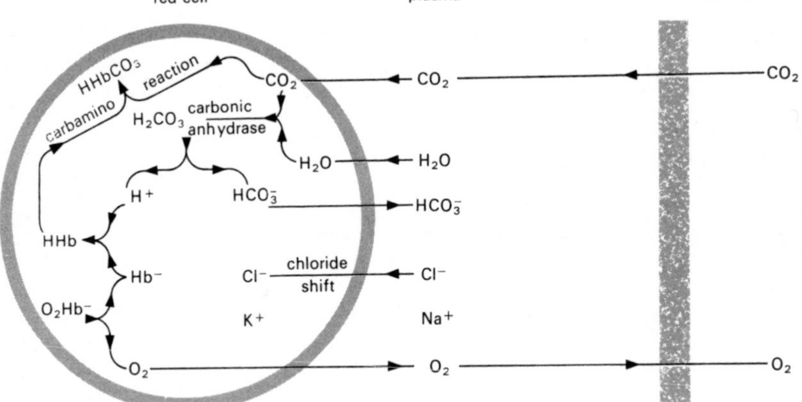

Figure 15.6c
A schematic summary of the chemical processes which occur when haemoglobin takes up oxygen in the lungs (above) and parts with its oxygen in the tissues (below). The chloride shift is represented as a movement of ions. The shaded part represents the capillary wall.
After Roughton, E. J. W. (1933) Physiol Rev. **15**, *293*

Let us now see how the pH of the blood, effectively that of the plasma, is related to the partial pressure of carbon dioxide in the alveolar air.

The equilibrium constant for the equilibrium

$$CO_2(aq) + H_2O(l) \rightleftharpoons H^+(aq) + HCO_3^-(aq)$$

is $$K_a = \frac{[H^+(aq)]_{eqm}[HCO_3^-(aq)]_{eqm}}{[CO_2(aq)]_{eqm}}$$

and $[CO_2(aq)]_{eqm} = bp_{CO_2eqm}$

where b is a solubility constant for carbon dioxide in blood plasma. It is defined as the number of millimoles of carbon dioxide dissolved in 1 cubic decimetre of plasma when $p_{CO_2} = 760$ mmHg and the temperature is 38 °C (the normal body temperature).

$$\text{Hence, } K_a = \frac{[H^+(aq)]_{eqm}[HCO_3^-(aq)]_{eqm}}{bp_{CO_2eqm}}$$

$$\therefore [H^+(aq)]_{eqm} = \frac{K_a\, bp_{CO_2eqm}}{[HCO_3^-(aq)]_{eqm}}$$

$$\therefore \mathrm{pH} = -\lg K_a - \lg \frac{bp_{CO_2eqm}}{[HCO_3^-(aq)]_{eqm}} \qquad (1)$$

This is a special form of the general equation relating to buffer solutions:

$$\mathrm{pH} = -\lg K_a - \lg \frac{[\text{acid}]_{eqm}}{[\text{base}]_{eqm}}$$

$bp_{CO_2eqm} = [CO_2(aq)]_{eqm}$ is the acid concentration, and $[HCO_3^-(aq)]_{eqm}$ is the base concentration in the equilibrium

$$H_2O(l) + CO_2(aq) \rightleftharpoons H^+(aq) + HCO_3^-(aq)$$

Doctors and biochemists refer to equation (1) as the *Henderson-Hasselbach* equation.

If the HCO_3^- ion concentration is expressed in millimole dm^{-3} and p_{CO_2} in mmHg, K_a has the value 7.9×10^{-7} mol dm^{-3} and b is 0.03 millimole dm^{-3}. In the plasma, the ratio

$$\frac{[CO_2(aq)]_{eqm}}{[HCO_3^-(aq)]_{eqm}} = \frac{bp_{CO_2eqm}}{[HCO_3^-(aq)]_{eqm}}$$

may be taken to be 0.05. Hence the normal pH value of plasma is 7.4. If the value of $[HCO_3^-(aq)]$ is 25.0 millimole dm^{-3}, verify for yourself that p_{CO_2eqm} in the alveolar air is 42 mmHg. In practice, we may take as standard the values $p_{CO_2eqm} = 40$ mmHg, pH = 7.4, and $[HCO_3^-(aq)]_{eqm} = 25.0$ millimole dm^{-3} although some variations from these figures are found even in healthy people.

Disturbances in the acid-base chemistry of the body

In a hospital biochemical laboratory the pH and p_{CO_2} of blood samples taken from patients suffering from respiratory diseases are regularly measured. If these two values are known, the $[HCO_3^-(aq)]$ can be calculated from the Henderson-Hasselbach equation but, as you will appreciate, repeated calculations of that nature would be time-wasting. Nomograms have therefore been constructed from the Henderson-Hasselbach equation. Figure 15.6d shows such a nomogram. If pH and p_{CO_2} are measured and a ruler laid across the nomogram so as to join the values found, the total carbon dioxide present in the plasma, the dissolved carbon dioxide, and the $[HCO_3^-(aq)]$ can be read off.

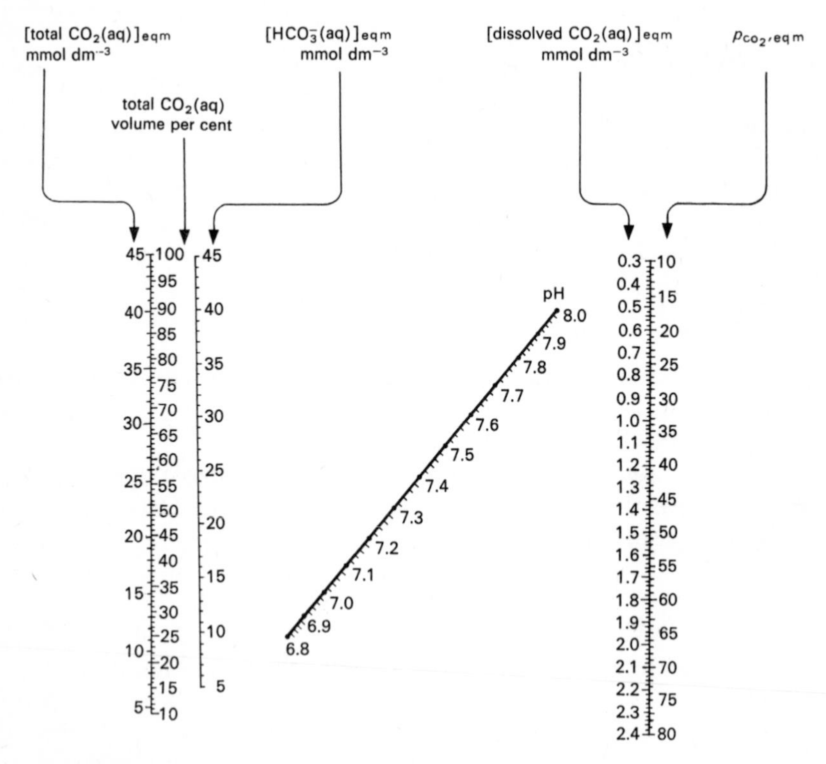

Figure 15.6d
A nomogram. *After McLean, F.* (*1938*) Physiol. Rev. **18**, *495.*

Blood analyses are also carried out in the operating theatre. When a patient is anaesthetized, it is often necessary for his breathing to be taken over by a mechanical system, or ventilator. In such cases, the anaesthetist must ensure that the gas mixture supplied to the patient is of the correct composition to maintain the p_{CO_2}, p_{O_2}, and pH at the normal values. Blood analyses in conjunction with a suitable nomogram assist him in this task. Nowadays, major heart surgery is commonplace and heart transplants have been performed. During such operations, the patient's breathing and circulation are taken over by a machine, the heart-lung machine, which must duplicate as closely as possible the normal function of the heart and lungs. In the heart-lung machine, blood from the patient is oxygenated and carbon dioxide removed from it, therefore constant pH and p_{CO_2} checks need to be made if the correct conditions are to be maintained.

An example is the case of a man suffering from severe bronchitis who was admitted to the intensive care unit of a London hospital. He was unconscious and blood samples showed the values $p_{CO_2} = 110$ mmHg and pH $= 7.31$. This man's disease prevented him from adequately ventilating his alveoli so that the partial pressure of carbon dioxide rose to the high value found. The buffer reserve of his blood had been exceeded and the plasma pH had fallen; this is a condition known as respiratory acidosis. The cure in this case is relatively straightforward, the patient's lungs are ventilated mechanically, after the insertion of a tube in his trachea, with an oxygen rich gas mixture. Whilst the ventilator maintains his breathing, the lung condition is treated with antibiotics until he is able to breathe normally again and maintain his own acid-base balance.

Another example is the use of the heart-lung machine. In figure 15.6e we see the variations in pH and p_{CO_2} in a patient undergoing heart surgery. From the start of the operation, the patient is artificially ventilated and the drop in p_{CO_2} with accompanying rise in pH after $1\frac{1}{4}$ hours is due to overventilation. The anaesthetist corrects this by altering the rate of ventilation and the composition of his gas mixture so that, at time $2\frac{1}{2}$ hours, the acid-base balance, or status, of the patient is restored. The time then approaches when the patient must be attached to the heart-lung machine. The machine is primed with 10 dm^3 of blood from the transfusion unit. Fifteen minutes before the patient is attached, the analysis of the blood in the machine shows pH $= 7.25$, $p_{CO_2} = 42$ mmHg. In order to match the acid-base status of the patient's own blood as closely as possible, two things are done. Firstly, the p_{CO_2} in the gases used to oxygenate the blood in the machine will be reduced. As a result, carbon dioxide dissolved in the blood will pass out of solution and hence, the HCO_3^- ion concentration will fall. To compensate for this, a calculated amount of sodium hydrogen carbonate is added to the blood. How many millimoles of sodium hydrogen carbonate would you add to the original 10 dm^{-3} of blood at pH $= 7.25$, $p_{CO_2} = 42$ mmHg, in order to stabilize at pH $= 7.45$ and $p_{CO_2} = 30$ mmHg?

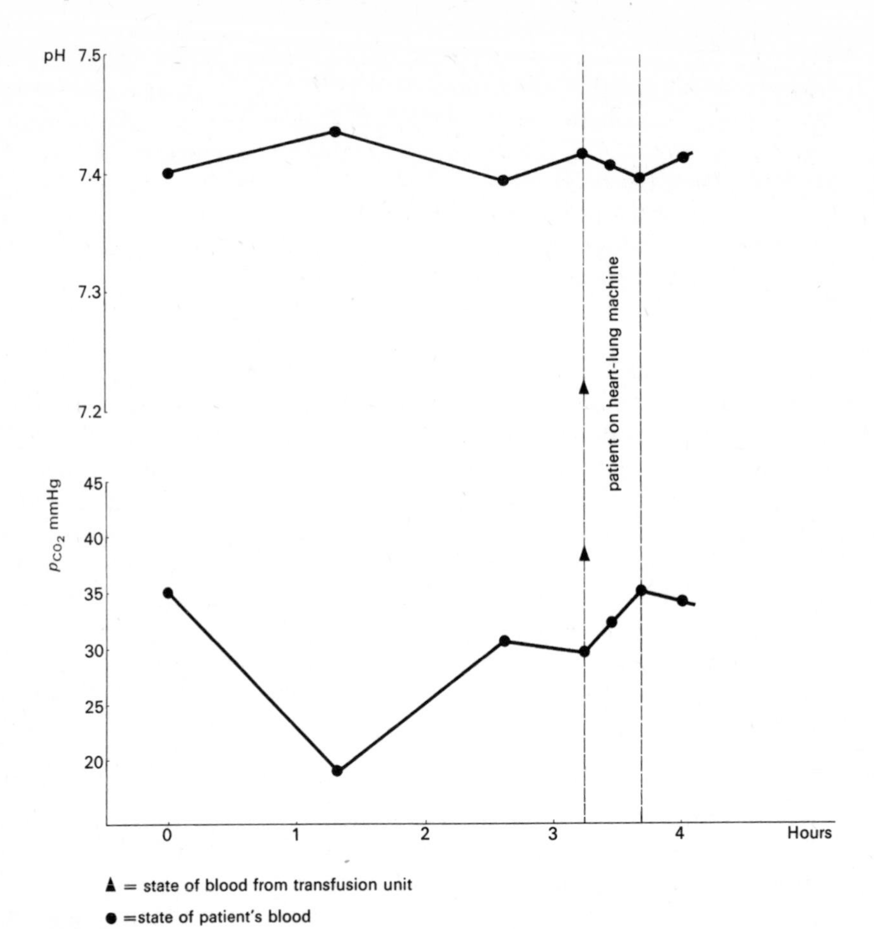

Figure 15.6e
The variation in pH and p_{CO_2} of the blood of a patient undergoing heart surgery.

Finally, let us consider briefly one other aspect of our subject. If the circulation of the blood fails to maintain an adequate supply of oxygen to the tissues, during an attack of coronary thrombosis, for example, incomplete oxidation of food in the muscles leads to an accumulation of lactic and pyruvic acids. These overcome the buffer reserve of the blood and produce a metabolic acidosis. To correct this, HCO_3^- is added to the patient's blood in the form of a solution of sodium hydrogen carbonate transfused into a vein. The amount required is calculated, after blood pH and p_{CO_2} measurements, from the Henderson-Hasselbach equation or, in practice, from a suitable nomogram.

In our discussions we have ignored the part played by the kidneys in regulating the acid-base chemistry of the body. You may wish to find out something about this for yourself.

Appendix I

Use of potentiometer to measure p.d.

A potentiometer method can be used to measure a potential difference accurately. The circuit is shown in figure 15.7.

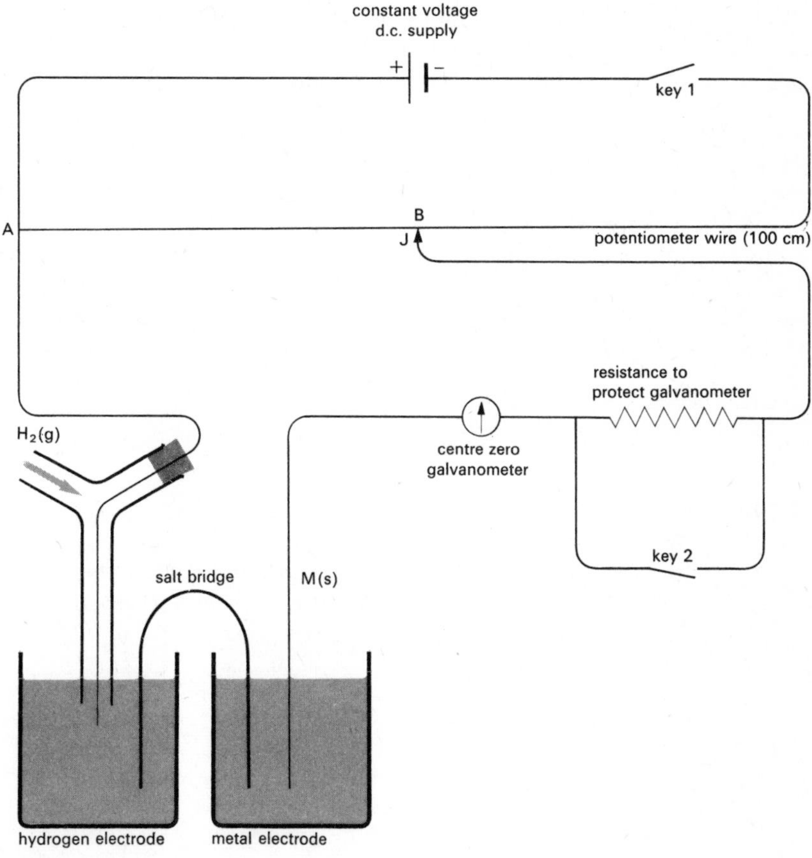

Figure 15.7

This method depends on opposing the current flowing in a circuit, of which the cell is a part, by a potential difference from another source of electricity, and adjusting this second p.d. until no current flows. The counter-voltage thus applied is equal to the e.m.f. of the cell being studied.

Procedure

With key 1 closed, the position of the sliding contact J is adjusted until the galvanometer reading is zero. This gives an approximate balance point only, because of the protective resistance in series with the galvanometer. A more accurate balance point can be obtained by shorting out the protective resistance (close key 2) and re-adjusting the sliding contact.

The fall in potential around the top part of the circuit in figure 15.7 occurs mainly along the potentiometer wire. When no higher accuracy is required, the e.m.f. of the cell being tested can be calculated easily from the known voltage of the d.c. supply. Thus, a 2 V accumulator could be used as the source of this current. The potential fall along the potentiometer wire is $\frac{2}{100} = 0.02$ V per cm. If the two parts of the circuit are in balance when the distance AB is 37.0 cm, the e.m.f. of the cell being tested is $37.0 \times 0.02 = 0.74$ V.

To obtain a more accurate value the potentiometer wire must be calibrated more carefully. This is done by replacing the test cell by a standard cell, the e.m.f. of which is accurately known for the temperature at which the measurement is being made. For example, a Weston standard cell* might be used which has an e.m.f. of 1.018 V at 20 °C. If, for this cell, the balance point is such that the distance AB is 49.4 cm, the e.m.f. of the cell is $\frac{37.0 \times 1.018}{49.4} = 0.726$ V.

Appendix II

Titration curves

On the following three pages there are some graphs showing the pH changes which take place when selected acids and bases react with each other. You may need to refer to these during class discussions.

* The cell diagram for the Weston cell is
$Cd(s) \mid CdSO_4(aq,satd.) \vdots Hg_2SO_4(aq,satd.) \mid Hg(l); E_{293} = +1.018$ V.

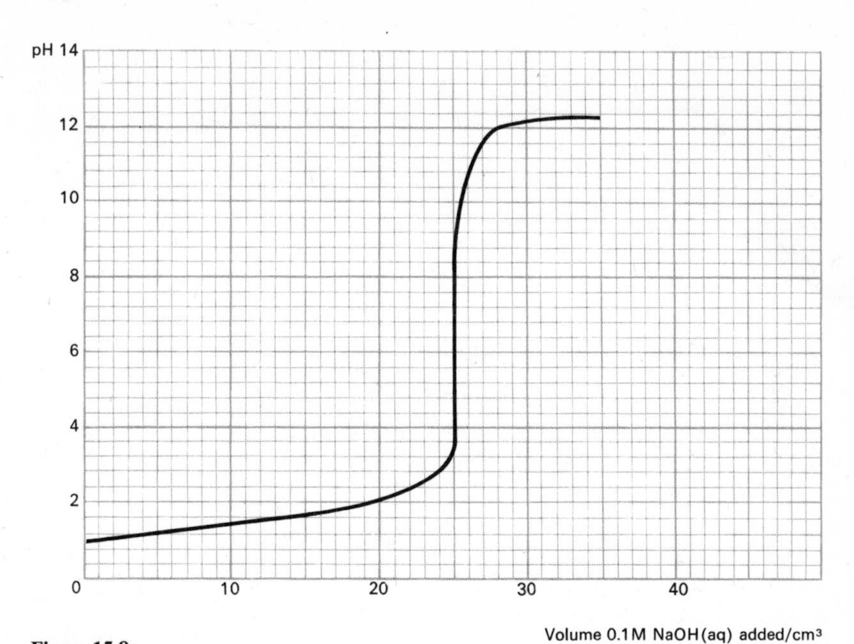

Figure 15.8a
pH changes during titration of 25 cm^3 0.1M HCl(aq) with 0.1M NaOH(aq).

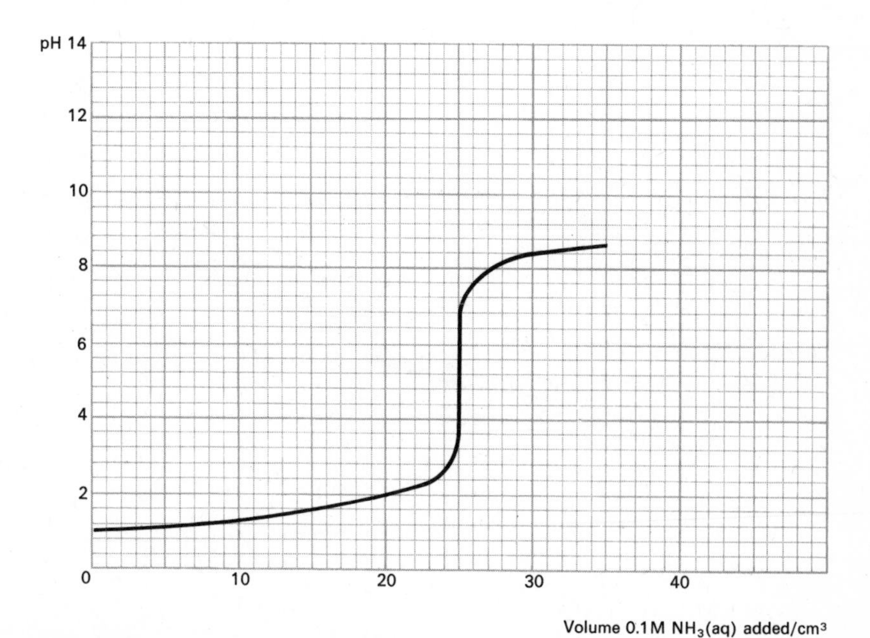

Figure 15.8b
pH changes during titration of 25 cm^3 0.1M HCl(aq) with 0.1M NH_3(aq).

Figure 15.8c
pH changes during titration of 25 cm³ 0.1M $CH_3CO_2H(aq)$ with 0.1M NaOH(aq).

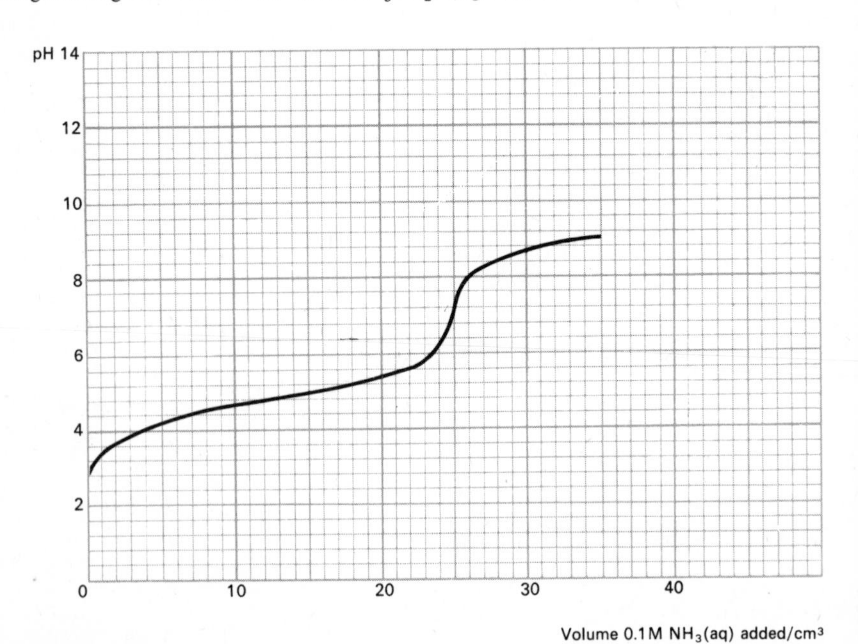

Figure 15.8d
pH changes during titration of 25 cm³ 0.1M $CH_3CO_2H(aq)$ with 0.1M $NH_3(aq)$.

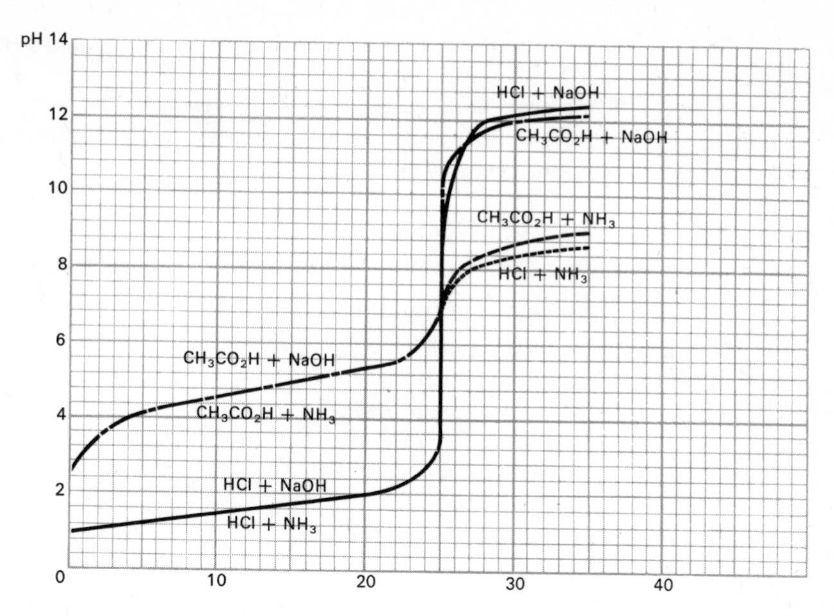

Figure 15.8e
Titration curves for strong and weak acids and bases.

Problems

* indicates that the *Book of Data* is required.

1 State which of the reactants is the oxidant and which is the reductant in each of the following reactions.

i $Fe(s) + Cu^{2+}(aq) \rightarrow Fe^{2+}(aq) + Cu(s)$

ii $Al(s) + 3H^{+}(aq) \rightarrow Al^{3+}(aq) + 1\frac{1}{2}H_2(g)$

iii $Zn(s) + Pb^{2+}(aq) \rightarrow Zn^{2+}(aq) + Pb(s)$

iv $2Fe^{3+}(aq) + Sn^{2+}(aq) \rightarrow 2Fe^{2+}(aq) + Sn^{4+}(aq)$

2 Consider the following reactions and information on four elements A, B, C, D (these are not the symbols for the elements):

a $A(s) + B^{2+}(aq) \rightarrow A^{2+}(aq) + B(s)$

b $A(s) + 2C^{+}(aq) \rightarrow A^{2+}(aq) + 2C(s)$

c $B(s) + 2C^{+}(aq) \rightarrow B^{2+}(aq) + 2C(s)$

d $D(s) + B^{2+}(aq) \rightarrow D^{2+}(aq) + B(s)$

e $D(s) + A^{2+}(aq)$ no action

i Are the elements likely to be all metals, all non-metals, or some of each? Briefly state the reason for your decision.

ii What reaction, if any, would you expect to take place between solid D and an aqueous solution of C^{+} ions?

iii Write two half-equations to represent the changes which occur in each of the reactions (a), (b), (c), and (d).

iv Arrange the elements in order of their relative tendency to form positive ions in aqueous solution, putting that with the greatest tendency first. Briefly state the reasons for your decision.

***3** Calculate the $E^\ominus$ value and state the terminal polarity of each of the following cells (assume a temperature of 25 °C and ionic concentration 1.0M).

i $Pt\,[H_2(g)]\,|\,2H^+(aq)\,\vdots\,Fe^{2+}(aq)\,|\,Fe(s)$
ii $Ni(s)\,|\,Ni^{2+}(aq)\,\vdots\,2H^+(aq)\,|\,[H_2(g)]\,Pt$
iii $Zn(s)\,|\,Zn^{2+}(aq)\,\vdots\,Ni^{2+}(aq)\,|\,Ni(s)$
iv $Al(s)\,|\,Al^{3+}(aq)\,\vdots\,Cr^{3+}(aq)\,|\,Cr(s)$

***4** $E^\ominus$ is +0.52 volt for the cell:
$Co(s)\,|\,Co^{2+}(aq)\,\vdots\,Cu^{2+}(aq)\,|\,Cu(s)$
Calculate the standard electrode potential for
$Co^{2+}(aq)\,|\,Co(s)$

***5** $E^\ominus$ is +1.61 volt for the cell:
$Zn(s)\,|\,Zn^{2+}(aq)\,\vdots\,Hg^{2+}(aq)\,|\,Hg(l)$
Calculate the standard electrode potential for
$Hg^{2+}(aq)\,|\,Hg(l)$

***6** For each of the following cells construct the two half equations and the whole equations to represent the changes which take place when the cell terminals are connected by a conductor.

i $Al(s)\,|\,Al^{3+}(aq)\,\vdots\,Sn^{2+}(aq)\,|\,Sn(s)$
ii $Ag(s)\,|\,Ag^+(aq)\,\vdots\,Pb^{2+}(aq)\,|\,Pb(s)$
iii $Pt\,[H_2(g)]\,|\,2H^+(aq)\,\vdots\,Mg^{2+}(aq)\,|\,Mg(s)$

***7** Put the following groups of ions in order of their *ability to oxidize*. Put the one with the greatest ability to oxidize first. (Assume that they are all of molar concentration.)

i $Cu^{2+}(aq)$ $Ag^+(aq)$ $Pb^{2+}(aq)$ $Cr^{3+}(aq)$
ii $Mg^{2+}(aq)$ $Zn^{2+}(aq)$ $Fe^{3+}(aq)$ $Sn^{2+}(aq)$

8 The redox potential of an electrode of a metal (M) was measured when it was in contact with solutions of its own ions at various molarities, using a standard hydrogen electrode. The results, at 25 °C, were:

Redox potential/volt	**Molarity of solution**
−0.286	0.5
−0.298	0.2
−0.307	0.1
−0.316	0.05
−0.327	0.02

i Plot a graph of redox potential against lg [ion].

ii From the graph determine the *standard* redox potential of the metal in an aqueous solution of its own ions. Explain how you arrive at your answer.

iii Determine the charge on the ions of the metal. Explain how you arrive at your answer.

9 To a solution of 10 cm^3 of 0.1M $AgNO_3(aq)$, 50 cm^3 of 0.1M potassium bromate, $KBrO_3(aq)$, solution were added. A piece of silver foil was put into the solution and its *electrode potential* at 25 °C was found to be +0.61 volt. Use the graph plotted from the results of experiment 15.1e to calculate:

i The concentration of aqueous silver ions in the final solution

ii The solubility product of silver bromate.

***10** Calculate the solubility of silver thiocyanate (AgCNS) in g dm^{-3} at 25 °C.

11 One cubic decimetre of a saturated solution of calcium oxalate contains 0.068 g CaC_2O_4 at 25 °C. Calculate the solubility product of calcium oxalate at this temperature.

***12** Calculate the solubility of iron(III) hydroxide in g dm^{-3} at 25 °C.

13 A certain metal (M) can exist in acidic aqueous solution with two oxidation numbers, one of which is +2 ($M^{2+}(aq)$). The following data give the electrode potentials at 25 °C of various aqueous mixtures of the two ions of the metal, M^{2+} and M^{x+} (x is greater than 2).
Relative concentrations

$[M^{2}(aq)]:[M^{x+}(aq)]$	***E*/volt**
8 : 1	0.113
4 : 1	0.122
1 : 2	0.149
1 : 5	0.161
1 : 10	0.170

By a graphical method determine $E^{\ominus}$ for the electrode, $M^{x+}(aq), M^{2+}(aq) \mid Pt$. What is the value of x? Explain how you arrive at your answers.

14 Classify the following systems as either *redox* or *acid/base* and to each of the components allocate the appropriate terms from amongst: $acid_1$, $acid_2$, $base_1$, $base_2$; $reductant_1$, $reductant_2$, $oxidant_1$, $oxidant_2$.

i $C_2H_5CO_2H(l)+H_2O(l) \rightleftharpoons C_2H_5CO_2^-(aq)+H_3O^+(aq)$
ii $Pb(s) + 2Ag^+(aq) \rightleftharpoons Pb^{2+}(aq) + 2Ag(s)$
iii $Mg(s)+2H^+(aq) \rightleftharpoons Mg^{2+}(aq)+H_2(g)$
iv $H_3O^+(aq)+OH^-(aq) \rightleftharpoons 2H_2O(l)$

15 Calculate the pH of the following solutions at 25 °C. In parts (*i*) to (*iv*) assume complete ionization.

i 0.2M HCl
ii 0.2M KOH
iii 0.125M HNO_3
iv A mixture of 75 cm^3 0.1M HCl and 25 cm^3 0.1M NaOH
v 0.1M bromoacetic acid (CH_2BrCO_2H; $K_a = 1.35 \times 10^{-3}$ mol dm^{-3})

***16** What is the concentration of formate ion (in mol dm^{-3}) in 0.01M formic acid solution at 25 °C?

17 In a 0.1M solution of a certain acid HA, $[A^-(aq)]_{eqm} = 1.3 \times 10^{-3}$ mol dm^{-3}; calculate K_a for the acid.

18 A 0.1M solution of a certain acid HA has a pH of 5.1; calculate K_a for the acid.

***19** Calculate the pH of an 0.001M solution of phenylammonium chloride at 25 °C. Assume that phenylammonium chloride ($C_6H_5NH_3Cl$) is fully ionized.

***20** Calculate the pH of a solution which is 0.1M with respect to propionic acid and 0.05M with respect to sodium propionate.

***21** In what proportions must 0.1M solutions of acetic acid and sodium acetate be mixed to obtain buffer solutions of

i pH 4.7
ii pH 4.4?

Topic 16
Some d-block elements

16.1 Introduction to d-block elements. Variable oxidation number

A definition of a d-block element is *one which forms some compounds in which there is an incomplete inner sub-shell of d-electrons.*

Looking at the first long period of the Periodic Table, potassium to krypton, this definition would include titanium to copper. Scandium and zinc, which are also in the gap between calcium and gallium, are not included in this definition.

Some of the properties of these elements are given in table 16.1a.

Element	**Sc**	**Ti**	**V**	**Cr**	**Mn**	**Fe**	**Co**	**Ni**	**Cu**	**Zn**
Electronic configuration (argon 'core' plus)	$3d^14s^2$	$3d^24s^2$	$3d^34s^2$	$3d^54s^1$	$3d^54s^2$	$3d^64s^2$	$3d^74s^2$	$3d^84s^2$	$3d^{10}4s^1$	$3d^{10}4s^2$
Atomic radius/nm	0.144	0.132	0.122	0.117	0.117	0.116	0.116	0.115	0.117	0.125
Oxidation numbers other than zero that occur (important ones are encircled)					⑦					
				⑥	6	6				
			⑤	5	5	5	5			
		④	4	4	4	4	4	4		
	③	3	3	③	3	③	③	3	3	
		2	2	2	②	②	②	②	②	②
		1	1	1	1	1	1	1	①	

Table 16.1a
(All oxidation numbers quoted are positive.)

Variable oxidation number

All the d-block elements can have more than one oxidation number (other than zero). As can be seen from table 16.1a, the highest oxidation number rises to 7 at manganese. The reason for the maximum oxidation numbers of potassium, calcium, scandium, titanium, vanadium, chromium, and manganese being 1, 2, 3, 4, 5, 6, and 7, is that if another electron is to be involved, it would have to come from the argon core of electrons. To do this would require very much more energy than to involve the outer electrons (in 3d and 4s) as can be seen from the ionization energies in figure 16.1a.

The most important oxidation number of these elements is that involving all the 3d and 4s electrons, that is, 4 for Ti, 5 for V, 6 for Cr, and 7 for Mn. After this, the increasing nuclear charge binds the nearer d electrons more firmly and so the most important (or one of the most important) oxidation number is that involving the more distant 4s electrons which are not so firmly bound, that is, 2 for Fe, Co, and Ni, and 1 for Cu. The stability of Mn(II) and Fe(III) can be attributed to the fact that the electronic structure of Mn^{2+} and Fe^{3+} is the same, $3d^5$. Just as a full electron shell has a particular stability, so a half filled shell, as in these ions, also has a particular stability, though less than that of the full shell.

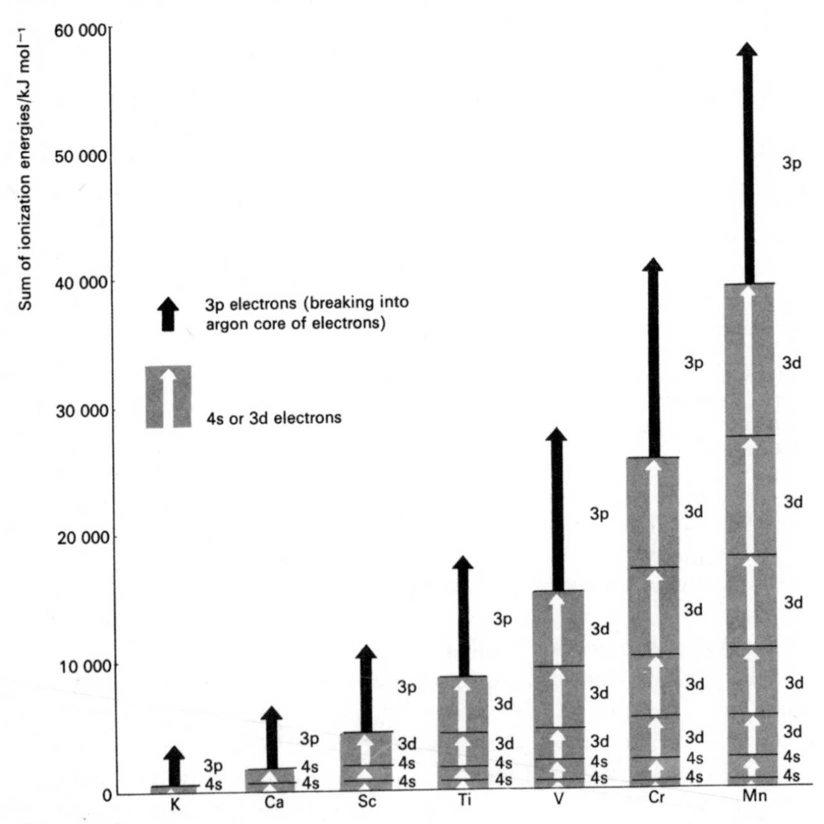

Figure 16.1a
Ionization energies of elements K to Mn.

Some of the other properties which the d-block elements have in common will be studied in this topic, but the first study will be of variable oxidation number.

$E^{\ominus}$ values in order of increasing positive (or decreasing negative) values are also in the order of decreasing tendency for the electrode system to release electrons, as shown in table 16.1b.

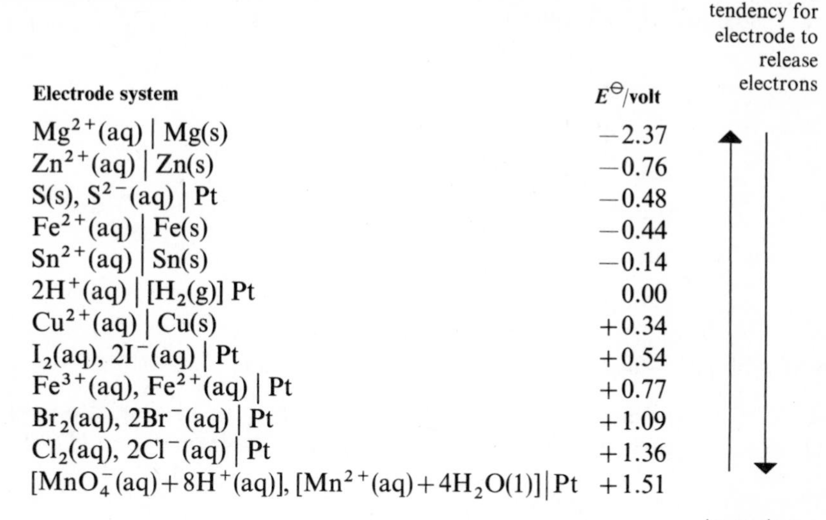

decreasing tendency for electrode to release electrons

Electrode system	$E^{\ominus}$**/volt**
$Mg^{2+}(aq) \mid Mg(s)$	−2.37
$Zn^{2+}(aq) \mid Zn(s)$	−0.76
$S(s), S^{2-}(aq) \mid Pt$	−0.48
$Fe^{2+}(aq) \mid Fe(s)$	−0.44
$Sn^{2+}(aq) \mid Sn(s)$	−0.14
$2H^{+}(aq) \mid [H_2(g)]\ Pt$	0.00
$Cu^{2+}(aq) \mid Cu(s)$	+0.34
$I_2(aq), 2I^{-}(aq) \mid Pt$	+0.54
$Fe^{3+}(aq), Fe^{2+}(aq) \mid Pt$	+0.77
$Br_2(aq), 2Br^{-}(aq) \mid Pt$	+1.09
$Cl_2(aq), 2Cl^{-}(aq) \mid Pt$	+1.36
$[MnO_4^{-}(aq)+8H^{+}(aq)], [Mn^{2+}(aq)+4H_2O(l)] \mid Pt$	+1.51

increasing tendency for electrode to release electrons

Table 16.1b

Hence if we link two electrode systems to form a voltaic cell, the system which is higher in the series will become the negative pole (transferring electrons to external circuit) and the system which is lower will become the positive pole. This is the same as saying that, as we go down the series, the oxidizing power of the oxidized forms in the electrodes increases and the reducing power of the reduced forms decreases. This can be illustrated by the $Fe^{3+}(aq)$, $Fe^{2+}(aq)$ and $I_2(aq)$, $2I^{-}(aq)$ reaction studied earlier. The order of these electrode systems is

$$I_2(aq), 2I^{-}(aq) \mid Pt;\ E^{\ominus} = +0.54\ V$$
$$Fe^{3+}(aq), Fe^{2+}(aq) \mid Pt;\ E^{\ominus} = +0.77\ V$$

In a cell made from these two electrodes, the $Fe^{3+}(aq)$, $Fe^{2+}(aq)$ system will be the positive pole. $Fe^{3+}(aq)$ is a better oxidizing agent than $I_2(aq)$, and $2I^{-}(aq)$ is a better reducing agent than $Fe^{2+}(aq)$. When the cell is working the reaction in the upper electrode goes from *right* to *left*, $2I^{-}(aq) \rightarrow I_2(aq)$, and that in the lower electrode from *left* to *right*, $Fe^{3+}(aq) \rightarrow Fe^{2+}(aq)$. This is the general situation for all reactions of this kind and can be summed up in a simple rule. For any pair of couples in the redox series, reaction will tend to go in such a way that taking the individual species in the couples in *anti-clockwise order, starting with the bottom left in the positions that they occupy in the series, gives the reactants and products of the possible reaction*. In the redox series we have

$I_2(aq), 2I^-(aq) \mid Pt$
above
$Fe^{3+}(aq), Fe^{2+}(aq) \mid Pt$

Start from $Fe^{3+}(aq)$ and proceed anti-clockwise

$$Fe^{3+}(aq) \xrightarrow{+e^-} Fe^{2+}(aq)$$
$$2I^-(aq) \xrightarrow{-2e^-} I_2(aq)$$

balance electron loss and gain

$$2Fe^{3+}(aq) \xrightarrow{+2e^-} 2Fe^{2+}(aq)$$
$$2I^-(aq) \xrightarrow{-2e^-} I_2(aq)$$

and add $2Fe^{3+}(aq) + 2I^-(aq) \rightarrow 2Fe^{2+}(aq) + I_2(aq)$

A more convenient form for prediction is to write the half-equations for the electrodes, rather than the electrode systems, in a diagram (figure 16.1b).

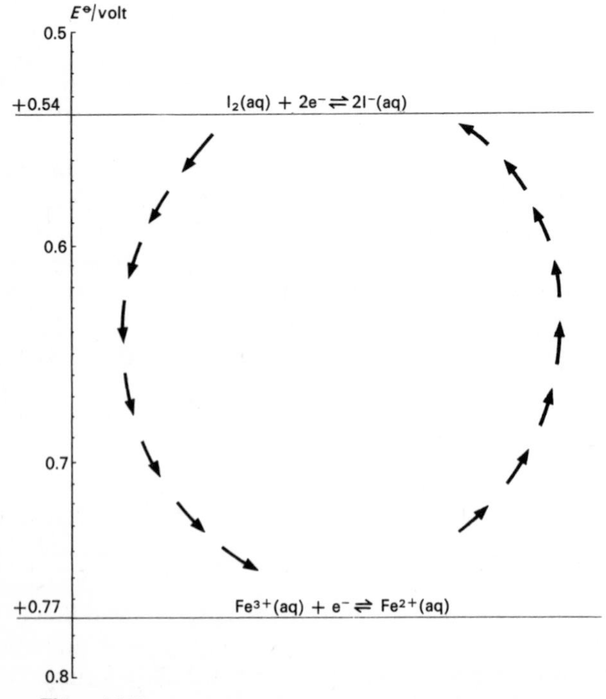

Figure 16.1b

Predictions made in this way refer strictly speaking only to standard conditions – molar solutions of reactants and products present at the same time and a fixed temperature. In other conditions, concentration or temperature effects may alter the equilibrium position (this is what we are really forecasting) of the opposing electrode systems and lead to a partial reversal of the change. If the difference in $E^\ominus$ values for the electrodes concerned is greater than 0.4 V, this is unlikely to happen.

Consider another example. Will potassium permanganate, in acid solution, be likely to oxidize hydrogen sulphide to sulphur? From the redox series we find the electrodes in the following order

$$[2H^+(aq)+S(s)], H_2S(aq) \mid Pt; E^\ominus = +0.14 \text{ V}$$
$$[MnO_4^-(aq)+8H^+(aq)], [Mn^{2+}(aq)+4H_2O(l)] \mid Pt; E^\ominus = +1.51 \text{ V}$$

and this can be written in a diagram as in figure 16.1c.

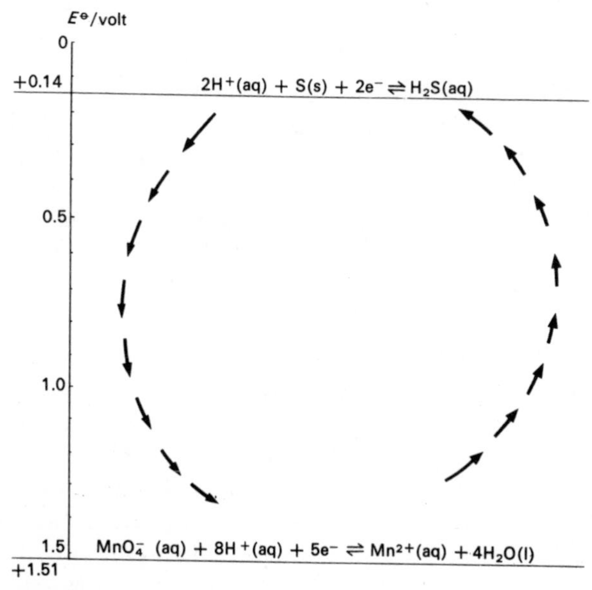

Figure 16.1c

The half-reactions should proceed as follows (the equations are more complicated since more than one species is involved in the oxidized and reduced electrode systems).

$$MnO_4^-(aq)+8H^+(aq) \xrightarrow{+5e^-} Mn^{2+}(aq)+4H_2O(l)$$
$$H_2S(aq) \xrightarrow{-2e^-} 2H^+(aq)+S(s)$$

Balance electron transfer

$$2MnO_4^-(aq) + 16H^+(aq) \xrightarrow{+10e^-} 2Mn^{2+}(aq) + 8H_2O(l)$$
$$5H_2S(aq) \xrightarrow{-10e^-} 10H^+(aq) + 5S(s)$$

and add

$$2MnO_4^-(aq) + 6H^+(aq) + 5H_2S(aq) \rightarrow 2Mn^{2+}(aq) + 8H_2O(l) + 5S(s)$$

The difference between the $E^\ominus$ values is considerable ($1.51 - 0.14 = 1.37$ V) so we should expect this reaction to proceed under all concentration conditions at ordinary temperatures. The difference in $E^\ominus$ values is, however, no guarantee that it will proceed quickly. In fact, this example is a reasonably fast reaction. Although we can use $E^\ominus$ values to tell us something about the position of equilibrium for a given change, they can never tell us how long it will take for this equilibrium to be attained. We have to experiment to find out the rate of a given reaction and, if it is slow, look for a catalyst if we want to make use of the reaction. A table of $E^\ominus$ values enables us to predict whether a search for a catalyst is worthwhile.

Experiment 16.1a

Investigation of the oxidation numbers of vanadium

Begin the experiment with a solution of vanadium(V). This is best made by dissolving ammonium metavanadate in acid. The ion present in solution under these conditions can be thought of as the vanadium(V) oxide ion, $VO_2^+(aq)$. Actually it polymerizes to a more complicated species. (In alkaline solution, the ion present can be thought of as $VO_3^-(aq)$, the vanadate ion, but it, too, polymerizes to a more complicated species.)

The other substances which you will need are:

		Species present	Made from
1	Fe(II)	$Fe^{2+}(aq)$	iron(II) ammonium sulphate
2	Fe(III)	$Fe^{3+}(aq)$	iron(III) sulphate
3	I(−I)	$I^-(aq)$	potassium iodide
4	I(O)	$I_3^-(aq)$	iodine dissolved in solution 3
5	Br(−I)	$Br^-(aq)$	potassium bromide
6	S(IV)	$SO_2(aq)$	sulphur dioxide in water
7	Cu(O)		solid copper powder
8	Zn(O)		solid zinc dust

Other species present in solution may be:

9	V(IV)	$VO^{2+}(aq)$	vanadium(IV) oxide ion
10	V(III)	$V^{3+}(aq)$	vanadium(III) ion
11	V(II)	$V^{2+}(aq)$	vanadium(II) ion
12	S(V)	$S_2O_6^{2-}(aq)$	dithionate ion
13	S(VI)	$SO_4^{2-}(aq)$	sulphate ion
14	Cu(II)	$Cu^{2+}(aq)$	copper(II) ion
15	Zn(II)	$Zn^{2+}(aq)$	zinc(II) ion
16	Br(O)	$Br_2(aq)$	bromine

The $E^{\ominus}$ values that you will require in this experiment are given in figure 16.1d.

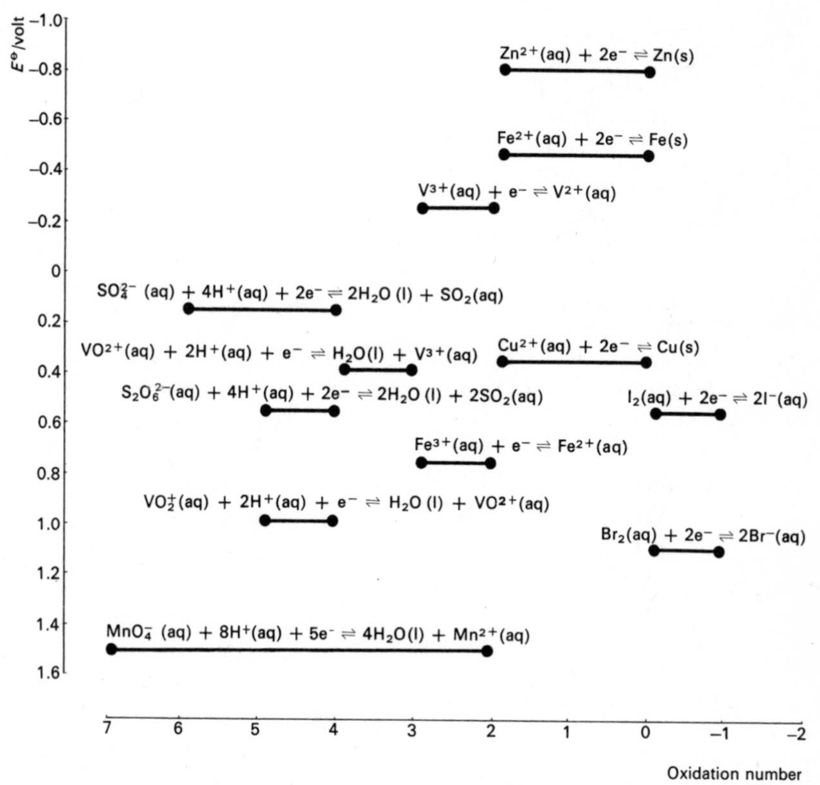

Figure 16.1d
Redox potentials at pH = 0.

Copy into your notebook and fill in the prediction columns of tables 16.1c and 16.1d given below. Then carry out the experiments and fill in the experimental results to see if your predictions are verified.

In each case add the other solution to the vanadium solution, or boil the solid with the vanadium solution and centrifuge or filter quickly.

You must be prepared to interpret your observations carefully. For instance, mixing a blue and a yellow solution will produce a green solution even if no chemical reaction has occurred. If iodine is produced in a reaction, the colour of other products of the reaction can only be seen if the iodine is converted to colourless iodide ions by addition of a solution of thiosulphate ions.

Substances to be mixed	Prediction	Experimental result
1 V(v)+Fe(II)		
2 V(v)+I(−I)		
3 V(v)+S(IV)		
4 V(v)+Cu(O)		
5 V(v)+Zn(O)		

Table 16.1c

What are the colours of the V(IV), V(III), and V(II) ions?

Substances to be mixed	Prediction	Experimental result
6 V(IV)+V(II)		
7 V(v) +V(III)		
8 V(IV)+V(III)		
9 V(IV)+Fe(III)		
10 V(III)+Fe(III)		
11 V(IV)+I(−I)		
12 V(v) +Br(−I)		
13 V(III)+I(O)		
14 V(IV)+I(O)		

Table 16.1d

If you have time, you may like to follow this investigation of vanadium with another one on manganese. The following are suitable instructions for such an investigation.

Experiment 16.1b

Investigation of the oxidation numbers of manganese

On the laboratory shelves are samples of Mn(II), probably as manganese(II) sulphate; Mn(IV), as manganese(IV) oxide; and manganese(VII), potassium permanganate. Mn(O) may also be available.

Arrange samples of these compounds and solutions of the soluble ions on a display card:

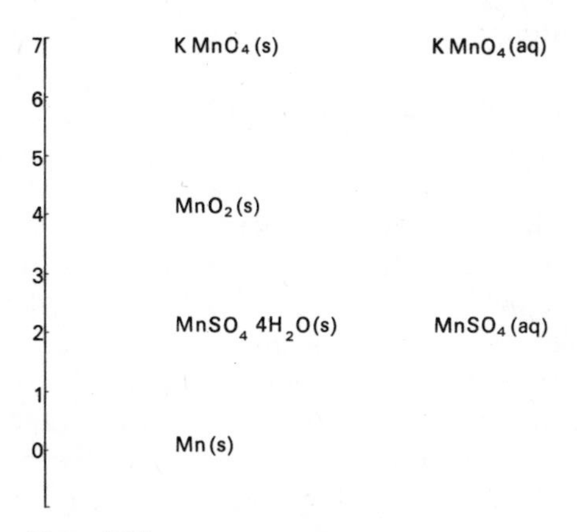

Figure 16.1e
Oxidation number display chart for manganese.

We will now attempt to obtain samples of other oxidation numbers in solution.

It is often found that a given oxidation number of an element can be obtained by reacting together compounds in which the element has higher and lower oxidation numbers.

Manganese(VI)

Manganese(VII) and manganese(IV) are available. Do redox potentials tell us anything about the likelihood of manganese(VI) being made from them?

In acid

$$2e^- + 2MnO_4^- \rightarrow 2MnO_4^{2-}; \quad E^\ominus = +0.56\ V$$
$$\text{Mn(VII)} \qquad \text{Mn(VI)}$$

$$2e^- + 4H^+ + MnO_4^{2-} \rightarrow MnO_2 + 2H_2O; \quad E^\ominus = +2.26\ V$$
$$\text{Mn(VI)} \qquad \text{Mn(IV)}$$

1 Will Mn(VII) and Mn(IV) react together under standard acid conditions to give Mn(VI)?

2 In view of the $E^\ominus$ values, do you think that altering the conditions from the standard state would alter the course of the reaction?

3 If the acid concentration is raised, what effect will it have?

In alkali

$2e^- + 2MnO_4^- \rightarrow 2MnO_4^{2-}$; $E^\ominus = +0.56$ V

$2e^- + 2H_2O + MnO_4^{2-} \rightarrow MnO_2 + 4OH^-$; $E^\ominus = +0.6$ V

4 Will Mn(VI) be formed under standard conditions?

5 Do you think that alteration of conditions will alter the course of the reaction?

6 What effect will raising the alkali concentration have?

Investigate this reaction by the following test-tube reactions. Take three 10 cm^3 portions of an 0.01M solution of $KMnO_4$. To one of these add about 5 cm^3 of dilute sulphuric acid and to another add about 5 cm^3 of dilute sodium hydroxide solution. Now add to each of the three tubes a little solid MnO_2. Shake each tube for a minute or so. In which tube does a reaction appear to have taken place? Check this by filtering each of the contents into clean tubes. The colour of the filtrate in the case where reaction has occurred is that of manganese(VI) in solution. Note the colour and preserve a small sample for the oxidation state chart. To another portion of the manganese(VI) solution add about 5 cm^3 of dilute sulphuric acid.

7 What happens?

8 What ion has been formed?

Add two or three pellets of potassium hydroxide and shake.

9 What happens?

10 What ion has been formed?

Put a sample of the solution of Mn(VI) on the display card.

Manganese(III)

Mn(II) and Mn(IV) are available.

In acid

$e^- + MnO_2 + 4H^+ \rightarrow Mn^{3+} + 2H_2O$; $E^\ominus = +0.95$ V

$e^- + Mn^{3+} \rightarrow Mn^{2+}$; $E^\ominus = +1.51$ V

11 Will Mn(II) and Mn(IV) react to give Mn(III) under standard acid conditions?

12 Will alteration of conditions alter the reaction?

13 If the acid concentration is raised, what effect will it have?

In alkali

$e^- + Mn(OH)_3 \rightarrow Mn(OH)_2 + OH^-$; $E^\ominus = +0.1$ V

$e^- + 2H_2O + MnO_2 \rightarrow Mn(OH)_3 + OH^-$; $E^\ominus = -0.2$ V

14 Will the reaction occur under standard alkaline conditions?

15 What effect will altering the alkali concentration have on the reaction?
16 Write down the equation for the reaction. Comment on the rate you might expect this reaction to have.

Add $\frac{1}{4}$ test-tube 2M sodium hydroxide to $\frac{1}{4}$ test-tube Mn(II) solution.
17 What happens?
18 On standing for a few minutes what happens?
19 What do you think has been formed?
20 How do you think it has been formed?

Another possibility might be to mix Mn(II) and Mn(VII).
21 When Mn(VII) oxidizes something, to what is it *normally* reduced?
In acid
$e^- + Mn^{3+} \rightarrow Mn^{2+}$; $E^\ominus > +1.51$ V
$5e^- + 8H^+ + MnO_4^- \rightarrow Mn^{2+} + 4H_2O$; $E^\ominus < +1.51$ V
22 Comment on the possibility of this reaction.
23 What effect will increasing the acid concentration have on the reaction?

Dissolve about 0.5 g of hydrated manganese(II) sulphate in about 2 cm^3 of dilute sulphuric acid. Add about 10 drops of concentrated sulphuric acid. Cool the tube under the tap and then add about five drops of 0.1M $KMnO_4$ solution. The resulting solution is a deep red in colour which is due to manganese(III). Save a small sample for the oxidation number chart and pour the remaining solution carefully into about 50 cm^3 of water, shake, and note the result. Find out if the same result is obtained by adding five drops of 0.1M $KMnO_4$ solution to a dilute acidified solution of $MnSO_4$ (say 0.5 g in 50 cm^3 water).
24 What happens when the Mn(III) solution is added to water?
25 Why do you think this happens?

Manganese (V)

Mn(IV) and Mn(VI) might react:
In acid
$e^- + 2H^+ + MnO_4^{2-} \rightarrow \underset{\text{Mn(V)}}{MnO_3^-} + H_2O$; $E^\ominus = +2.0$ V
$e^- + 2H^+ + MnO_3^- \rightarrow MnO_2 + H_2O$; $E^\ominus = +2.5$ V
26 Can MnO_3^- be made in this way?
27 What effect will decreasing the pH have?

In alkali
$e^- + H_2O + MnO_4^{2-} \rightarrow MnO_3^- + 2OH^-$; $E^\ominus = +0.34$ V
$e^- + H_2O + MnO_3^- \rightarrow MnO_2 + 2OH^-$; $E^\ominus = +0.84$ V
28 Can MnO_3^- be made in this way?
29 What effect will increasing the pH have?

You may be shown a method of preparing Mn(V) from Mn(VII).

30 Mn(VII) has been reduced to Mn(V):

$2e^- + MnO_4^- + H_2O \rightarrow MnO_3^- + 2OH^-$

What has been oxidized to what?

? $\rightarrow$? + $2e^-$

To five drops of 0.1M potassium permanganate solution, add two pellets of potassium hydroxide.

31 Is there any colour change?

32 To what oxidation number has the Mn(VII) been reduced? Write a half equation.

33 What has been oxidized to what to bring this reduction about? Write a half equation.

Background reading

Structure and uses of three metals – titanium, chromium, and copper

Titanium

Titanium has a hexagonal close packed crystal structure, which changes at 882.5 °C to body-centred cubic, and the metal has the relatively high melting point of about 1670 °C. Its density is low (4.5 – only one half that of copper), and its high strength-to-weight ratio together with its excellent corrosion resistance led to extensive research and development of titanium production facilities in the 1950s.

Titanium ores are abundant, although the cost of extraction of the metal is high. This is because the only commercial method of producing titanium is by reduction of titanium(IV) chloride, with molten magnesium in an atmosphere of argon. The metal owes its remarkable resistance to corrosion to the presence of a thin, tenacious oxide film, and the application of this metal is usually where its high price is offset by the need for a substance which will be reliable under conditions which would cause corrosion in all other known constructional materials.

The alloys of titanium fall into three groups: (1) the all-alpha alloys, such as those of titanium and aluminium which are hexagonal in structure, have good strength and toughness but relatively poor forming characteristics; (2) the all-beta alloys, such as those of titanium and copper which are body-centred cubic in structure, have much better forming characteristics, have good strength both hot and cold, but, in general, have a lower strength-to-weight ratio; and (3) the alpha + beta alloys, such as those of titanium and manganese which represent a compromise in properties between those of (1) and (2).

Chromium

Chromium has a body-centred cubic crystal structure and melts at 1903 °C. Its density (7.1) is low compared with that of its potential competitors, tungsten (19.3) and molybdenum (10.2), and the metal has a high resistance to oxidation which, as in the case of titanium, is due to the presence of a thin oxide film upon its surface.

Unless in a state of very high purity, chromium is extremely brittle, so that forging it and using it structurally are not possible unless special precautions, such as melting it by electric arc in vacuum, are taken to keep contamination (especially by nitrogen) to a minimum. Ductile chromium must never contain more than 0.02 per cent nitrogen. One familiar use of the pure metal is in 'chromium plating' where a very thin 'flash' of chromium (2×10^{-5} cm thick) is deposited upon a more ductile substrate of copper and nickel (about 2×10^{-3} cm thick).

Copper

Copper has a cubic close packed crystal structure and melts at 1083 °C. Since it is a relatively 'noble' metal ('native' copper is occasionally encountered), it is very easy to extract from its ores, which is why it has been known from the earliest times (Bronze Age). Copper owes its corrosion resistance to its chemical nature, and not to the presence of a protective surface film.

Copper admirably exemplifies 'typical' metallic properties – it is malleable, ductile, and has high electrical and thermal conductivity. Although most metals find their major industrial use in the form of alloys, this is not true of copper. This metal is used most extensively as an electric conductor and in radiators, both of which uses require a high purity of metal.

Alloying copper with zinc produces the brasses. These were originally developed partly because they were cheaper than unalloyed copper, but they also show an increased strength and better forming characteristics. Adding tin to copper produces bronze, which is an alloy which can show high strength, corrosion resistance, and is also readily cast into a mould. Both copper and its alloys are coloured, which makes them aesthetically attractive for architectural purposes, and they often acquire an attractive green patina with age.

16.2 **Complex ions**

In Topic 15 we saw that acid-base equilibria were competitions for protons between bases, that is, between molecules or ions with unshared pair(s) of electrons. A dynamic equilibrium is set up in, for instance, a saturated solution of hydrogen chloride in water in a closed vessel. The bases $Cl^-(aq)$ and $H_2O(l)$ are competing for a proton:

$$Cl^-(aq) + H^+ \rightleftharpoons HCl(aq)$$
$$H_2O(l) + H^+ \rightleftharpoons H_3O^+(aq)$$

When the proton becomes attached to the base, a dative covalent bond is formed, the base supplying both electrons:

```
      ..                    ..+
      O                     O
     / \      + H+ →       /|\
    H   H                 H H H

  ..              ..
 :Cl:-  + H+ → H—Cl:
  ..              ..
```

In this case the water wins, and most of the protons are present as $H_3O^+(aq)$.

Other cations can be attracted towards a source of electrons and so hydration occurs with, for instance, Cu^{2+}, when the positive ion is attracted towards the electrons on the oxygen atoms:

$$Cu^{2+}(s) + 4H_2O(l) \rightleftharpoons Cu(H_2O)_4^{2+}(aq)$$

Similarly to H_3O^+, a dative covalent bond is formed with electrons going into the readily available electron levels on the copper ion (3d, 4s, and 4p). *Four* electron donor atoms are associated with the cation. This is possible because the cation is very much larger than a proton.

The ion formed in such a way is called a *complex ion*: it is a metal ion surrounded by a number of oppositely charged ions or neutral molecules called *ligands*. The metal ion and the ligands can exist separately.

Water is probably the commonest ligand. The hydration energy can be calculated and it can be regarded as the avidity with which ions are hydrated.

$$M^{z+}(s) + mH_2O(l) \rightarrow M(H_2O)_m^{z+}(aq)$$

The strength of the bonds formed when various ions are hydrated can then be compared by comparing the values of the hydration energies.

Ion	Radius/nm	Heat of hydration/kJ mol^{-1}
Li^+	0.060	469
Na^+	0.097	414
K^+	0.133	351
Be^{2+}	0.035	2385
Mg^{2+}	0.066	1828
Ca^{2+}	0.099	1510
Sr^{2+}	0.112	1356
Ba^{2+}	0.135	1243
Al^{3+}	0.051	4456

Table 16.2a
Heats of hydration of some ions

It can be seen that stronger hydration occurs for smaller ions and for those of higher charge. The ligand is more strongly attracted to a higher charge; and this charge is more 'concentrated' in a smaller ion.

So far we have only considered how strongly various ions are hydrated; what happens when two ligands compete for a cation? This is analogous to the situation:

$$HCl(g) + H_2O(l) \rightleftharpoons H_3O^+ + Cl^-(aq)$$

Which base wins? In this case the equilibrium is entirely to the right showing that H_2O is a stronger base than Cl^-: or that HCl is a stronger acid than H_3O^+, that is, HCl provides protons more readily than does H_3O^+.

For the reaction:

$$Ag^+(aq) + 2NH_3(aq) \rightleftharpoons Ag(NH_3)_2^+(aq)$$

the equilibrium constant is 10^7 which shows that the equilibrium is to the right; NH_3 wins against H_2O in the competition for Ag^+. As this equilibrium constant tells us about the stability of a complex, it is called the *stability constant*. It is difficult to predict stability constants, but relatively easy to determine them experimentally.

Experiment 16.2a

Experimental investigation of some complexes

Procedure

a Put five or six drops of 0.5M copper(II) sulphate solution in a test-tube.
1 What complex ion is present?
2 What is its colour?

b Add ten drops of concentrated hydrochloric acid drop by drop.
3 What is the colour of the solution now?
4 What ligands do you think are now present in the complex ion?

c Keep half of the solution for (d). Pour the other half into a test-tube half full of water.
5 What colour is the solution now?
6 What complex ion is now present?
7 In view of what happened in (b), why do you think this reaction occurred?

d To the solution kept from (c), add concentrated ammonia solution drop by drop till there is no further colour change. Save this solution for (f).
8 What colour is the solution now?
9 What ligands do you think are now present in the complex ion?
10 Qualitatively compare the relative stabilities of the various complex ions formed during these experiments: $Cu(H_2O)_4^{2+}$; $CuCl_4^{2-}$; $Cu(NH_3)_4^{2+}$.
11 Are your results consistent with the stability constants in table 16.2b?

Ligand		lg K_1	lg K_2	lg K_3	lg K_4	lg K
Cl^-	chloride	2.8	1.6	0.49	0.73	5.6
NH_3	ammonia	4.3	3.6	3.04	2.3	13.2
$C_6H_4(CO_2^-)(OH)$	2-hydroxybenzoate	10.6	6.3			16.9
$C_6H_4(OH)(OH)$	1,2-dihydroxybenzene	17.0	8.0			25.0
EDTA	(ethylenediamine-tetra-acetic acid)	18.8				18.8

Table 16.2b
Stability constants for complexes of copper(II)

K_1 is the equilibrium constant for: $Cu(H_2O)_4^{2+}(aq) + Cl^-(aq) \rightleftharpoons Cu(H_2O)_3Cl^+(aq) + H_2O(l)$
K_2 is the equilibrium constant for: $Cu(H_2O)_3Cl^+(aq) + Cl^-(aq) \rightleftharpoons Cu(H_2O)_2Cl_2(aq) + H_2O(l)$
K_3 is the equilibrium constant for: $Cu(H_2O)_2Cl_2(aq) + Cl^-(aq) \rightleftharpoons Cu(H_2O)Cl_3^-(aq) + H_2O(l)$
K_4 is the equilibrium constant for: $Cu(H_2O)Cl_3^-(aq) + Cl^-(aq) \rightleftharpoons CuCl_4^{2-}(aq) + H_2O(l)$
K is the equilibrium constant for: $Cu(H_2O)_4^{2+}(aq) + 4Cl^-(aq) \rightleftharpoons CuCl_4^{2-}(aq) + 4H_2O(l)$

e To four or five drops of $Cu(H_2O)_4^{2+}(aq)$ in a test-tube add a solution of the ligand EDTA until there is no further colour change.

12 What colour is the EDTA–Cu(II) complex?

13 If EDTA solution were added to the solution obtained in (d), what do you think would happen? (Use the stability constants in table 16.2b to make your prediction.)

f Add EDTA solution drop by drop to the solution obtained in (d) until there is no further colour change.

14 Was your prediction in question 13 correct?

g Put five or six drops of copper(II) solution in each of four test-tubes. To the first test-tube add EDTA solution drop by drop until there is no further colour change. To the second similarly add ammonia solution; to the third, sodium 2-hydroxybenzoate solution; and to the fourth, 1,2-dihydroxybenzene solution in 0.5M sodium hydroxide. Note that the last solution contains sodium hydroxide.

15 What are the colours of the four complexes?

16 You are going to carry out an experiment in which first a solution of ammonia, then one of sodium 2-hydroxybenzoate, then EDTA, then 1,2-dihydroxybenzene are added in turn to a solution of copper(II) ions. From the stability constants in table 16.2b, predict what will happen.

h Check your predictions by adding 8–10 drops of concentrated ammonia solution to 4–5 drops of copper(II) solution, followed by 10–12 drops of sodium 2-hydroxybenzoate solution, 5–6 drops of EDTA solution, and 10–15 drops of 1,2-dihydroxybenzene solution in turn. Add the solutions drop by drop noting the colour of the solution.

The 2-hydroxybenzoate ion, $C_6H_4(CO_2^-)(OH)$ 1,2-dihydroxybenzene, $C_6H_4(OH)_2$

and EDTA, ethylenediaminetetra-acetic acid,

$$(HO_2CCH_2)_2N{-}CH_2{-}CH_2{-}N(CH_2CO_2H)_2$$

are *polydentate* ligands. That is, they can form more than one link with the metal ion. Thus in solutions in which there is an excess of ligand present, the predominating species containing the metal are:

$Cu(H_2O)_4^{2+}$ monodentate ligands
$Cu(NH_3)_4^{2+}$
$CuCl_4^{2-}$

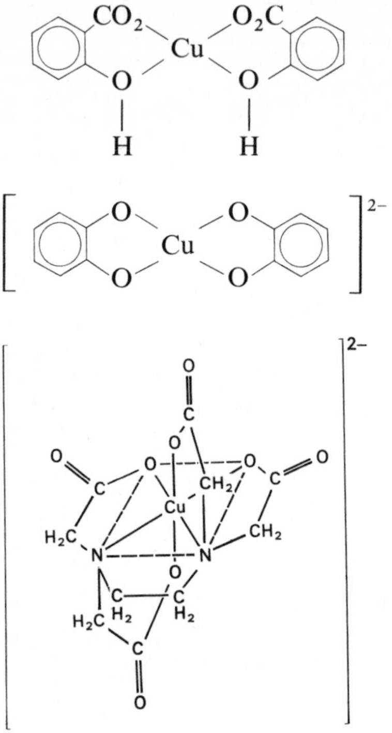

bidentate ligands

hexadentate ligand

Figure 16.2a

Experiment 16.2b

Investigation of the stoichiometry of the Ni(II)–EDTA complex ion

The stoichiometry of a complex can be found by the method of continuous variation, as long as only one species is formed.* Solutions are made up containing nickel(II) and EDTA in molar proportions 0:10, 1:9, 2:8, 3:7, etc. If the complex is $Ni(EDTA)^{2-}$ then the highest concentration of complex would be in the 5:5 mixture and the others would be more dilute. Hence if the concentration of complex can be determined in each mixture, the stoichiometry of the complex ion can be found.

Many complex ions, including the Ni(II)-EDTA ion, are coloured, and the concentration of a coloured substance can be found by using a colorimeter (previously described at the end of Topic 14). The absolute concentration of the complex is not required, only a measurement which is proportional to the concentration. Thus a meter reading, m, proportional to I can be measured

* This method can be used if more than one complex ion is formed, but the treatment is more complicated than the simple one indicated here.

for each mixture and then a graph of $\lg\left(\frac{I_o}{I}\right)$, on the y axis, against the molar proportions drawn. If the complex is $Ni(EDTA)^{2-}$ the graph will have the form shown in figure 16.2b.

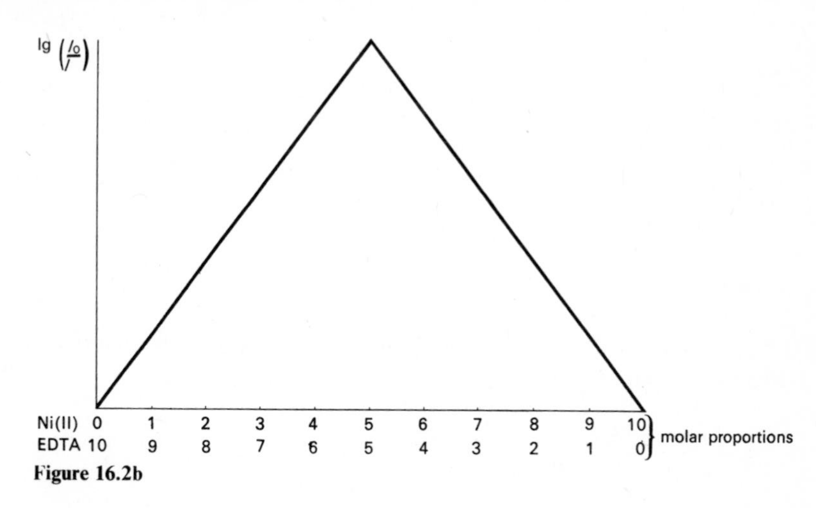

Figure 16.2b

In eleven tubes which fit the colorimeter make up mixtures:

1	0 cm^3 0.05M Ni(II) solution	+ 10 cm^3 EDTA solution
2	1 cm^3	+ 9 cm^3
3	2 cm^3	+ 8 cm^3
4	3 cm^3	+ 7 cm^3
5	4 cm^3	+ 6 cm^3
6	5 cm^3	+ 5 cm^3
7	6 cm^3	+ 4 cm^3
8	7 cm^3	+ 3 cm^3
9	8 cm^3	+ 2 cm^3
10	9 cm^3	+ 1 cm^3
11	10 cm^3	+ 0 cm^3

Choose the most suitable filter and then find the meter reading, m, for each mixture, adjusting the colorimeter to give $m_o = 1$ with a tube of water in place before each measurement.

Plot a graph of $\lg\left(\frac{I_o}{I}\right)$ against the molar proportions of EDTA and Ni(II) and hence find the stoichiometry of the complex.

Stereochemistry of complexes

Some interesting instances of stereoisomerism occur with complex ions. The hexamminechromium(II) ion is octahedral (figure 16.2c).

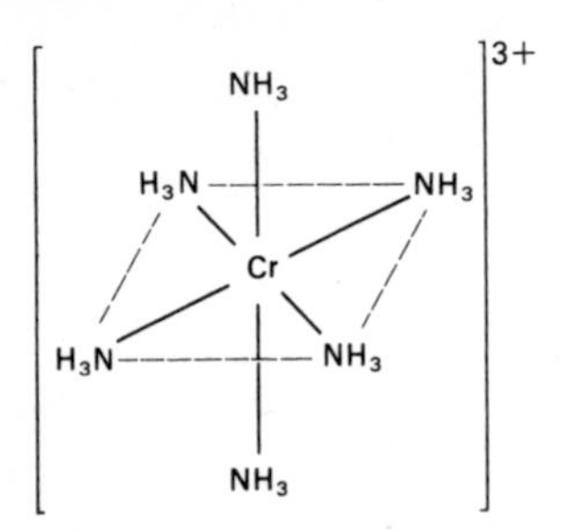

Figure 16.2c

If the six NH_3 ligands are replaced by three bidentate ligands then two isomers having the formula Cr en_3^{3+}, are possible:

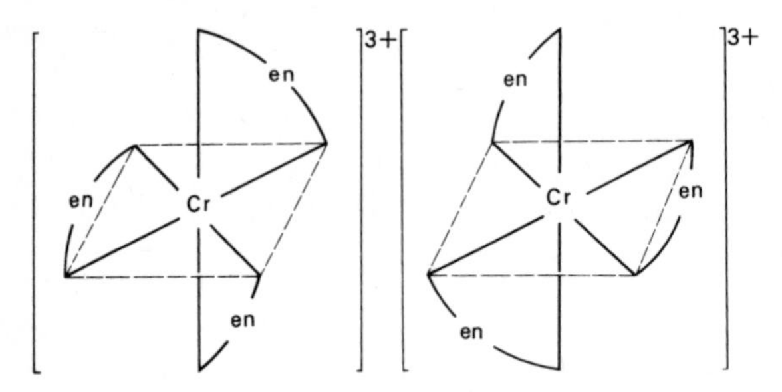

Figure 16.2d

where en = 1,2-diaminoethane: $NH_2CH_2CH_2H_2N$.
As these two isomers are mirror images one of the other, each is optically active. If there are two bidentate and two monodentate ligands, then three isomers occur, having the formula $(CrCl_2en_2)^+$, two of which are optically active:

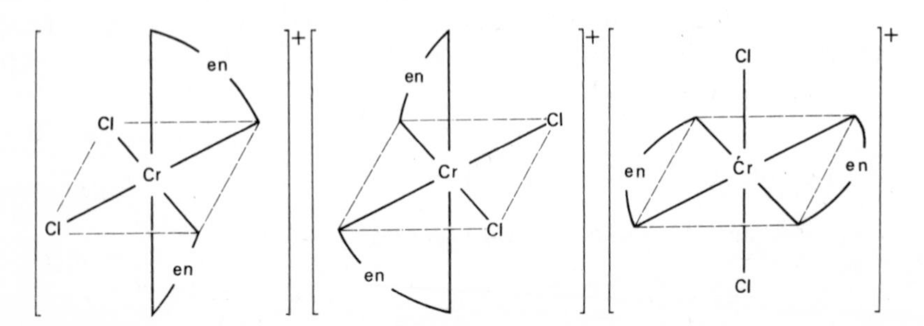

Figure 16.2e

A very interesting situation exists with the salt having the formula $Cr(H_2O)_6Cl_3$. Three isomers are known. The first, which is violet, has the formula

$$[Cr(H_2O)_6]^{3+}\ [Cl_3]^{3-}$$

The second, which is light green, has the formula

$$[CrCl(H_2O)_5]^{2+}\ [Cl_2]^{2-},H_2O$$

and the third, which is dark green, has the formula

$$[CrCl_2(H_2O)_4]^{+}\ [Cl]^{-},2H_2O$$

How many moles of silver chloride will be precipitated if excess silver ions are added to a solution containing 1 mole of

1 $[Cr(H_2O)_6]^{3+}\ [Cl_3]^{3-}$
2 $[CrCl(H_2O)_5]^{2+}\ [Cl_2]^{2-},H_2O$
3 $[CrCl_2(H_2O)_4]^{+}\ [Cl]^{-},2H_2O$?

Background reading

Uses of complexing agents

Gravimetric analysis

There are several reasons why it may be useful to precipitate a metal as a complex. Metal complexes often have very low solubilities and the error due to ions remaining in solution is very low. The metal is also part of a complex of high molecular weight which reduces the effect of errors in weighing and those due to loss of precipitate. But one of the most important features of this kind of precipitant is that, by adjusting the pH of the solution, it can often be made specific for only one metal. Dimethylglyoxime, for instance, in dilute ammonia solution, forms complexes with many metals but all are very soluble except for the nickel complex which precipitates.

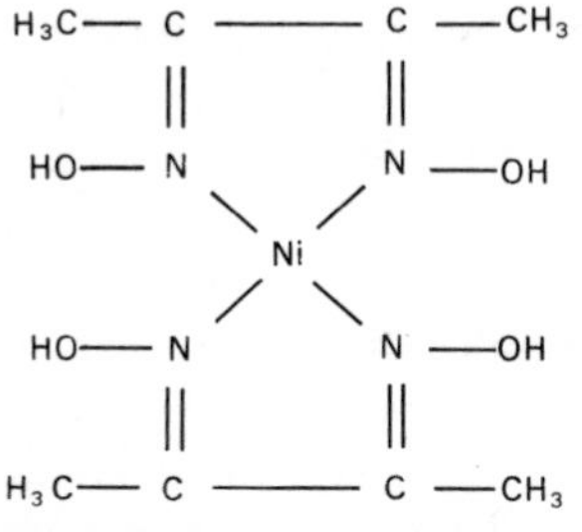

```
H3C— C ———— C —CH3
     ||      ||
HO— N        N —OH
      \     /
        Ni
      /     \
HO— N        N —OH
     ||      ||
H3C— C ———— C —CH3
```

Figure 16.2f
The nickel-dimethylglyoxime complex.

Colorimetric analysis

Although the hydrated ions of a metal may be colourless or absorb to only a small extent in the visible region of the spectrum, other complexes of the metal may absorb strongly and therefore very small concentrations of the metal can be determined colorimetrically. As the complex absorbs at a characteristic wavelength, either by using a filter or by means of a monochromator, that is, a prism or diffraction grating, the concentration of one metal ion in the presence of many others can be found.

The method finds extensive use in industry because it is both rapid and reasonably accurate, and is easily adapted to many of the routine determinations carried out in a quality control laboratory. For example, thioglycollic acid can be used to check that the iron content of a commercial sample of sodium carbonate is below 0.001 per cent. The thioglycollic acid-iron complex has a strong red colour and the colour produced by the sample can be compared with that produced by a standard sample of sodium carbonate which has an iron content of exactly 0.001 per cent.

Complexometric analysis

This is the titration of a metal ion with a suitable complexing agent. EDTA (ethylenediaminetetra-acetic acid), as its disodium salt, is probably most widely used.

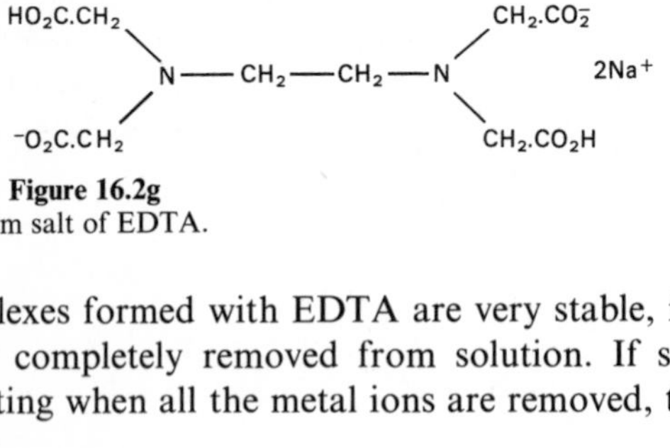

$HO_2C.CH_2$ and $^{-}O_2C.CH_2$ on N—CH_2—CH_2—N bearing $CH_2.CO_2^{-}$ and $CH_2.CO_2H$; $2Na^{+}$

Figure 16.2g
Disodium salt of EDTA.

Complexes formed with EDTA are very stable, i.e. almost all metal ions are nearly completely removed from solution. If some means is available for indicating when all the metal ions are removed, then EDTA can be used as a titrant.

Other complexing agents can be used as indicators. They must form strongly coloured complexes with the metal ion and the complex must be stable enough to be formed whenever a very small concentration of free metal ion is available. The metal-indicator complex must, of course, be less stable than the metal-EDTA complex so that when EDTA is in slight excess, all of the other complex will have disappeared. Many dyestuffs are used as indicators such as Eriochrome Black for magnesium, calcium, and manganese, Pyrocatechol Violet for bismuth and thorium, and Murexide for calcium, nickel, and copper.

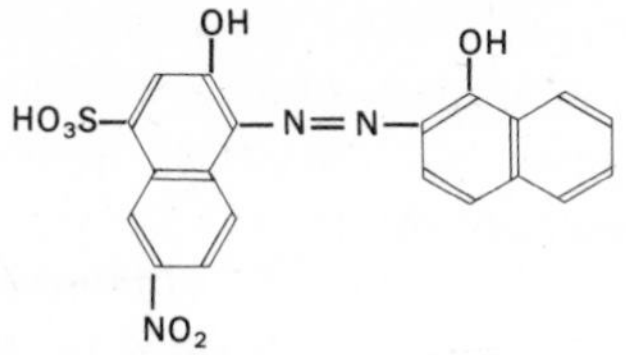

Figure 16.2h
Eriochrome Black T.

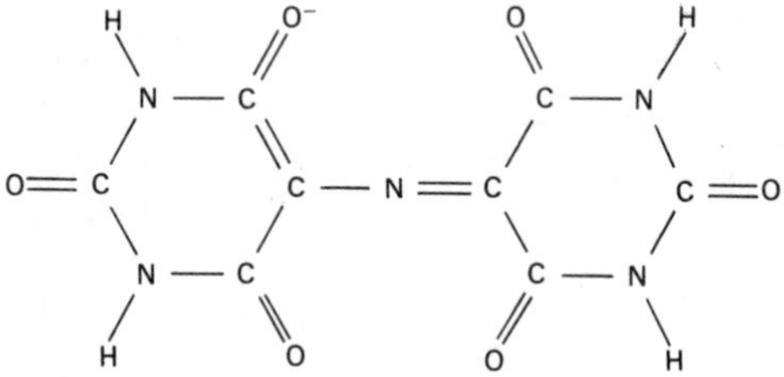

Figure 16.2j
Murexide ion.

Complexing agents for water softening

Hard water contains calcium or magnesium ions which form insoluble compounds with soap (the 'scum' left when the bath water is drained away). Many substances will form soluble complexes with calcium and magnesium ions, removing them from solution and 'softening' the water. Few have been used as water softeners mainly because of their expense, but the polyphosphates, especially sodium pyrophosphate ($Na_4P_2O_7$) and the cyclophosphates ('meta' phosphates), are available commercially, and are used in detergents. An example is Calgon, manufactured by Albright and Wilson.

Solvent extraction

If a solute is added to two immiscible solvents, A and B, in contact with each other, the solute distributes itself between the two and an equilibrium is set up between the solute molecules in solvent A and the solute molecules in solvent B.

$$\text{solute in A} \rightleftharpoons \text{solute in B}$$

The partition coefficient, $\dfrac{[\text{solute in A}]_{\text{eqm}}}{[\text{solute in B}]_{\text{eqm}}}$ is constant for a particular system at a particular temperature.

The partition coefficients of many metal complexes between water and organic solvents are such that when an aqueous solution of the complex is shaken with an organic solvent, the complex passes almost completely from the aqueous phase to the organic phase. In this way, a metal ion can be 'extracted' from an aqueous solution. By choosing suitable complexing agents and organic solvents and by controlling the pH of the aqueous solution, it is possible to separate one particular metal ion from a solution containing many metal ions. The complexes are usually strongly coloured and this allows the concentration in the organic layer to be determined colorimetrically. In this way a metal ion can be extracted from a mixture and its concentration in the mixture determined.

Dithizone (diphenyl thiocarbazone) forms complexes with many metals, and these complexes are soluble in organic solvents such as trichloromethane.

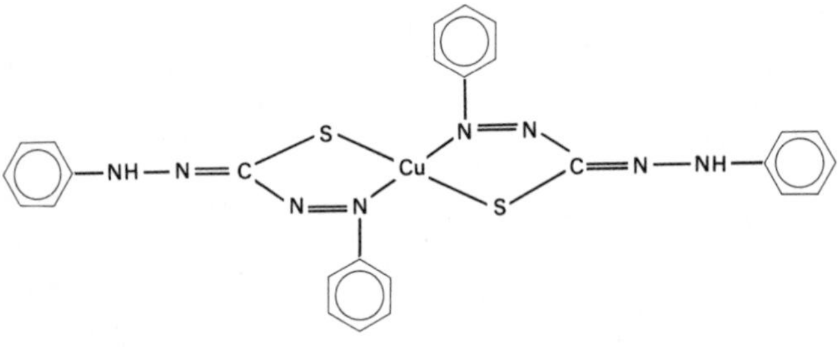

Figure 16.2k
The dithizone-copper complex.

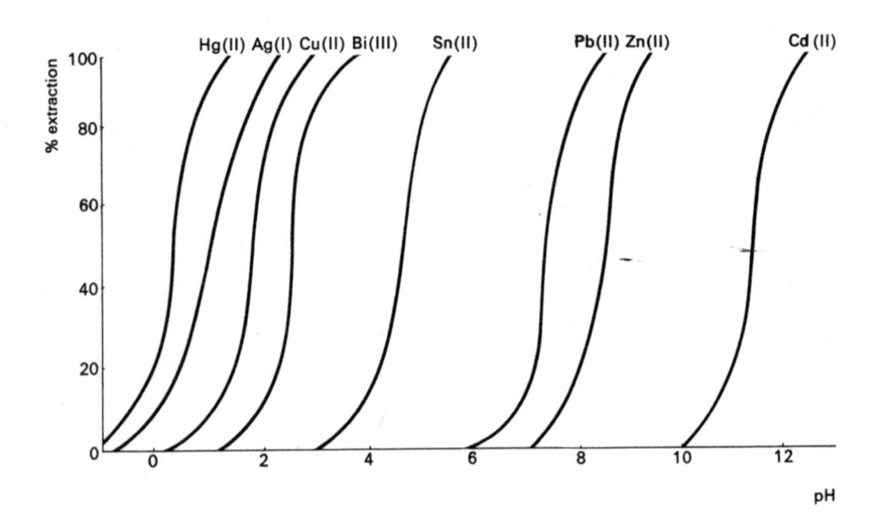

Figure 16.21

Figure 16.21 shows how the pH may be adjusted to enable certain ions to be separated from the rest.

Suppose that all the ions shown are present in aqueous solution and it is required to separate Sn(II). The dithizonates are prepared and extracted with chloroform from a solution of pH 5.5; Hg(II), Ag(I), Cu(II), Bi(III), and Sn(II) dithizonates will be extracted. On back-extracting with an aqueous solution of pH 3.0, only Sn(II) will pass into the aqueous solution.

The electrodeposition of metals

The fact that metals can be deposited onto a cathode by electrolysis of suitable salt solutions has formed the basis of many industrial processes. Examples are the refining of copper, nickel plating onto steel to prevent corrosion, the preparation of rare metals, and the manufacture of silver plate.

In all these processes, it is important to obtain a metallic deposit which is uniform and does not flake off either during electrolysis or during subsequent use. An electrolytic solution containing a simple salt of the metal normally gives a deposit which usually consists of non-adherent coarse particles. The addition of a complexing agent to the electrolytic solution often improves the deposit. An example is the use of cyanide in silver plating.

$$Ag^{+}(aq) + 2CN^{-}(aq) \rightleftharpoons Ag(CN)_2^{-}(aq)$$

The factors which control the nature of the metal deposit are complex and not yet fully understood, but high current densities and low non-uniform ion concentrations should be avoided. The presence of colloidal organic matter such as gelatin, camphor, or casein, is often helpful, perhaps by preventing high current densities near the surface of the cathode. Metal complexes are often colloidal, but probably a more important reason is that equilibrium is set up between the complex and free metal ions. This helps to avoid local variation in ion concentration near the cathode.

Complexes also change the electrode potential by reducing the metal ion concentration. Complexes can thus be used to control electrode potentials allowing metals to be selectively deposited. Electrode potentials can also be so adjusted that two metals are deposited simultaneously, a process useful in the manufacture of alloys.

Table 16.2c illustrates how the potentials of zinc in 0.1M $ZnSO_4$ and copper in 0.1M $CuSO_4$ can be brought close to each other by adding cyanide.

	With no KCN	In 0.2M KCN	In 1.0M KCN
Zn	−0.816 V	−1.033 V	−1.231 V
Cu	+0.292 V	−0.611 V	−1.169 V

Table 16.2c
Electrode potentials of zinc and copper

The extraction of metals from ores

Complexing agents can also be used in the extraction of metals from their ores. A good example is the use of cyanide in the extraction of gold from gold-bearing quartz. The powdered quartz is agitated with a solution of sodium cyanide when the gold passes slowly into solution:

$$4Au(s) + 8CN^-(aq) + 2H_2O(l) + O_2(g) \rightarrow 4Au(CN)_2^-(aq) + 4OH^-(aq)$$

The gold is then precipitated by adding zinc dust to the solution and the excess of zinc dust is removed by dissolving in dilute acid.

With sulphide ores (e.g. argentite)

$$Ag_2S(s) + 4CN^-(aq) \rightleftharpoons 2Ag(CN)_2^-(aq) + S^{2-}(aq)$$

With chloride ores (e.g. horn silver)

$$AgCl(s) + 2CN^-(aq) \rightleftharpoons Ag(CN)_2^-(aq) + Cl^-(aq)$$

Again zinc dust is used to precipitate metallic silver and the excess of zinc removed by dilute acid.

Dyes and pigments

Metal complexes are often very strongly coloured, very stable, and sometimes very insoluble in most solvents. These properties make them useful as dyes and pigments in such materials as plastics, paints, and printing inks. Complexes of iron, copper, chromium, cobalt, and nickel have been used in this way, giving a large range of colours.

One of the most important group of metal complexes used as pigments is the phthalocyanines. In 1928, a researcher at Scottish Dyes Ltd (now part of ICI) noticed that during the manufacture of phthalimide from phthalic anhydride and ammonia using iron vessels, an impurity was formed which had a very strong blue colour. This was iron phthalocyanine; it was separated and found to have very good properties as a pigment. Later, the copper derivative was found to be an even better blue pigment and this is now marketed under the name Monastral Blue.

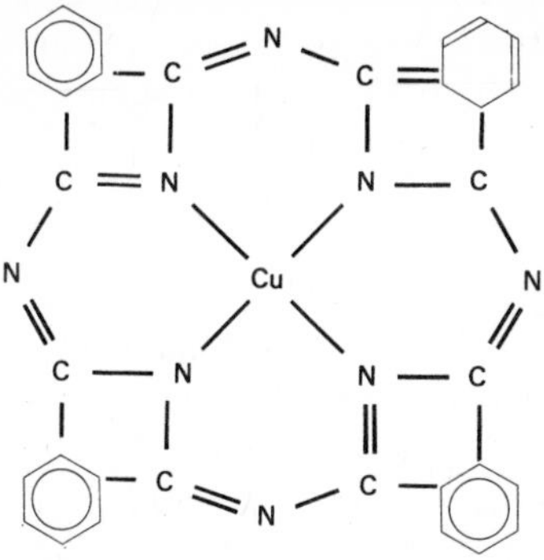

Figure 16.2m
Monastral Blue.

Complexing agents in biology and medicine

Metal complexes are quite common in biological systems. They are similar to the complexes discussed previously except that the ligands usually have a very high molecular weight and consist mostly of protein. Even the phthalocyanine molecules are quite small in comparison.

Iron, magnesium, and copper are the commonest metals found in these complexes but other metals such as cobalt, zinc, manganese, molybdenum, and vanadium are also encountered.

Some of these complexes are discussed in the section on trace elements at the end of this topic.

16.3 **d-Block elements in catalysis**

Experiment 16.3

A study of the catalytic action of d-block ions on the reaction between iodide and persulphate ions

Iodide ions are oxidized by persulphate ions to iodine

$$S_2O_8^{2-}(aq) + 2I^-(aq) \rightarrow 2SO_4^{2-}(aq) + I_2(aq)$$

and various metal ions will catalyse this reaction. A convenient way in which to measure the rate of this reaction is to add a fixed volume of sodium thiosulphate solution to the reaction mixture. This reacts with the iodine formed in the reaction as follows:

$$I_2(aq) + 2S_2O_3^{2-}(aq) \rightarrow 2I^-(aq) + S_4O_6^{2-}(aq)$$

When the thiosulphate has been used up, free iodine is produced, and if some starch solution has also been added, a deep blue colour will be produced. The time that it takes for the blue colour to appear is inversely proportional to the average rate of reaction during this time, and so, whether or not an ion catalyses the reaction can be found by comparing the times for the reaction with and without the metal ion.

Procedure

Place 10 cm^3 of iodide solution, 10 cm^3 of thiosulphate solution, and 5 cm^3 of starch solution in a 150 cm^3 flask, and 20 cm^3 of persulphate solution in a boiling tube. Add the persulphate solution to the flask and start the stopclock. Note the time taken for the blue colour to develop. Repeat the experiment adding three drops of a solution of one of the ions listed below. The ions can be shared out amongst the class so that the whole range is tested.

Cr(III)	Mn(II)	Fe(II)	Co(II)	Ni(II)	Cu(II)
Cr(VI)	Mn(VII)	Fe(III)			

Compare the times from the various experiments with and without metal ion and so find out if the addition of the metal ion affects the rate of the reaction. In order to compare the times some idea of the errors involved must be known. If two times are

1 130 ± 30 seconds, and
2 80 ± 20 seconds

then although inspection at 80 and 130 seconds might suggest that the reaction in (2) was nearly twice as fast as that in (1), the errors indicate that (1) could be as low as 100 seconds and that (2) could be as high as 100 seconds so the rate in the two experiments may be the same. Comparison of several results will enable you to gain some idea of the errors, and, indeed, an average of several experimental results will give better figures with which to decide whether the rate is being affected.

Questions

1 Which ions catalyse the reaction?
2 What sort of reaction is being catalysed? (Precipitation, redox, neutralization, etc.)
3 What is being transferred in the reaction?
4 How might the catalyst (an ion of a metal in the d-block) help this reaction?
5 Can you find out if your suggestion is possible by looking up $E^{\ominus}$ values?
6 Can you test your suggestions practically?

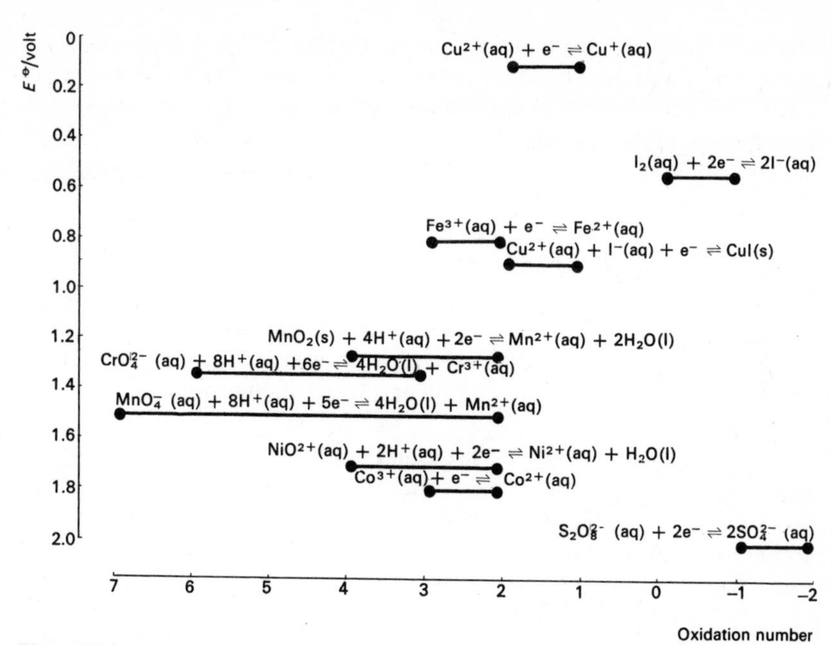

Figure 16.3a
Redox potentials at pH = 0.

Background reading

Trace elements

Many elements are found in combined form in living tissue, both plant and animal, in very small amounts. Early workers who were unable to measure the concentrations of these elements with the methods then available referred to such elements as 'trace elements'. The term is still used today although now we are able to measure the low concentrations in which they occur with great accuracy.

Examples are given in table 16.3.

	Cu	Mo	Co	Zn	Mn
Liver	24.9	3.2	0.18	55	1.68
Kidney	17.3	1.6	0.23	55	0.93
Brain	17.5	0.14		14	0.34
Muscle		0.14		54	0.09

Table 16.3
Trace elements in man (parts per million)

A number of trace elements have been shown to be essential to the enzyme mechanisms and metabolic processes in the body. However, some appear to be inert in the sense that, so far, their presence has not been linked to any vital function in plants or animals. Some of these elements, such as aluminium, silver, lead, gold, bismuth, tin, titanium, and gallium, may simply be acquired as a result of contact between the organism and its environment. An example of this sort of thing is the accumulation of ^{90}Sr in animals and man which occurred during the period when nuclear bombs were tested in the atmosphere. Another is the build up of pesticide residues, particularly organochlorine compounds in the bodies of animals and fish. Because so little is known about the possible physiological effects of such traces, society ought to exercise extreme caution before sanctioning any form of environmental pollution.

The study of trace elements in animals began about a century ago and was placed on a firm quantitative footing with the development of atomic emission spectroscopy. As you will appreciate from a reading of earlier topics, atomic emission spectroscopy is capable of quantitative assessment of elements in very low concentrations and is, therefore, an ideal tool in trace element studies.

A major impetus to the study of trace elements was provided by developments in nuclear physics during the 1950s. In the first place, the contamination of the atmosphere by test fall-out made a knowledge of the distribution of stable and radioactive elements in the body essential if an assessment of safe levels was to be made. Secondly, the increasing availability of radioactive isotopes manufactured in atomic piles enabled tracer techniques to be developed so that the routes by which certain trace elements reached organs of the body via the food chains could be established. Extensive spectrographic investigations and radioactive tracer studies undertaken at this time have established:

1 The normal limits of concentration of trace elements in plants and animals.

2 The influence of age, environment, and disease on these concentrations.

They have also indicated which elements are likely to prove to have biological significance, and which are merely acquired from the environment. This latter point has stimulated a great deal of research into the modes of action of elements previously regarded as inert. You will appreciate that this is a field of activity in which chemistry and biology have no boundary between them and that it is one in which a great deal of work is waiting to be done.

Let us now consider the role of some trace elements in more detail.

Copper

Copper is essential for life. Areas of the earth which have no available copper, such as certain parts of Australia, are deserts. What properties does copper have that make it so important? First, it forms very stable complexes with nitrogen-containing compounds such as amino acids and proteins. Second, it is a particularly good catalyst; and, third, it can exist in two oxidation states other than zero and can be readily converted from one to the other. Oxygen from the atmosphere can oxidize copper(I) to copper(II).

As with many trace elements, an excess of copper is toxic, but most animals and plants have a satisfactory supply of copper; its deficiency is very rare. Copper is present in several enzymes, and without it, these enzymes are not available to catalyse various important processes. Energy is obtained by oxidation of food in many stages. The last of these stages in all animals and in most plants is catalysed by the enzyme, cytochrome oxidase, which contains copper. Without copper this terminal reaction cannot occur and the whole chain is stopped. As this terminal oxidase is completely inhibited by cyanides, all animals and plants are very sensitive to these.

Another copper-containing enzyme, amine oxidase, catalyses the oxidation of the amino acid lysine, eventually to produce desmosine. This substance provides cross-links in elastin, which is the main protein in the wall of the aorta and other blood vessels, giving them their elasticity and stability. Without copper, desmosine is not formed and the walls of these blood vessels are weakened. This may result in the rupture of the blood vessels.

One of the most interesting biologically active substances that contain copper is haemocyanin. This blue substance is present in the blood of snails, octopuses, and scorpions and serves as an oxygen carrier. The blood of most animals contains red haemoglobin as the oxygen carrier. Snails, octopuses, and scorpions have virtually the only true blue blood that exists. Blood containing haemocyanin carries less oxygen than blood containing haemoglobin as the oxygen carrier. It is probable that initially all animals used haemocyanin as the blood oxygen carrier, but that higher animals turned to haemoglobin in the course of evolution, whereas the molluscs and arthropods continued to make use of haemocyanin which is adequate for their requirements.

Another enzyme which contains copper is tyrosinase. Amongst other functions, this enzyme controls the oxidation of the amino acid tyrosine and other phenols to melanin, the black pigment responsible for the colour of man's skin. It is also involved in the browning of apples, potatoes, and other fruit.

Copper is stored in higher animals as the protein ceruloplasmin in the blood serum. This protein may also act as an enzyme in the synthesis of haemoglobin. Certainly severe copper deficiency leads to a reduction in the rate of synthesis of haemoglobin and this leads to anaemia.

Iron

An adult man weighing 70 kg contains between 4 and 5 g of iron. Of this amount, it has been estimated that 60–70 per cent occurs in the haemoglobin of the red blood cells. The chemical mechanism by which iron is incorporated in the haemoglobin molecule is not yet fully understood, although the process is known to take place in the bone marrow. Iron is conveyed to the marrow in the form of an iron(III) complex with a protein, transferrin, which is found in blood plasma. Once in the marrow, it is reduced to iron(II) and appears in that form in haemoglobin.

The structural unit of haemoglobin which contains iron(II) is as shown in figure 16.3b.

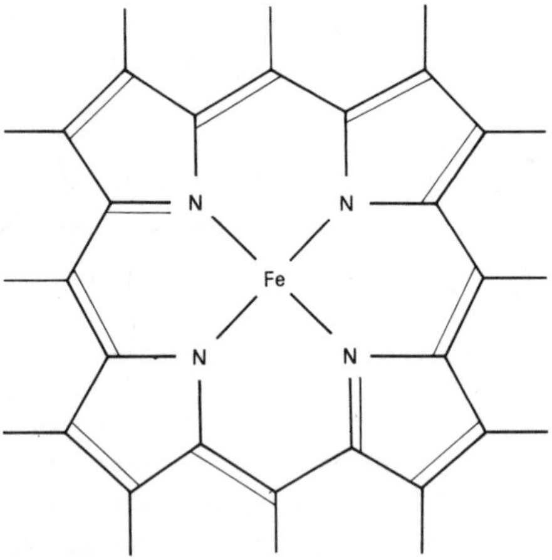

Figure 16.3b
The basic porphyrin unit of haemoglobin containing four-coordinated ions.

The metal atom is surrounded by a planar ring of four nitrogen atoms, that is, it is four-coordinated. Further coordination with an oxygen molecule is possible and by this means haemoglobin is responsible for the transport of oxygen in the blood. Deficiency of iron gives rise to one form of anaemia, a condition characterized by fatigue, shortness of breath, palpitation, and general malaise. The symptoms can be corrected by addition of iron to the diet and by doses of iron(II) sulphate.

Cobalt

A serious form of anaemia exists which is not due to iron deficiency but to a faulty development of the red corpuscle. This 'pernicious anaemia' occurs as a result of a failure on the part of the liver to supply a chemical compound essential to the production of red cells. In 1948, it was shown that this was a cobalt compound, which we now know as vitamin B_{12}, and which contains 4 per cent of the metal.

Zinc

Zinc occurs in quite high concentrations in many organs of the body. It appears to play an important role in the male sex organs and fluids, although the nature of the compounds of zinc present in the testes and semen is as yet unknown. Human spermatozoa contain 1990 parts per million of zinc, and zinc deficiency in animals has been shown to lead to atrophy of the testes.

Manganese

The function of this element in the human body has not yet been established with certainty but manganese deficiency in animals is associated with crippling deformities of the skeleton, and with sterility.

Trace elements and enzymes

Small amounts of d-block metals have been shown to be essential for the functioning of enzymes in the body. There are two categories; in one, the metal forms a compound with the enzyme itself to give a *prosthetic group* or *metallo-enzyme* and, in the second, no compound is formed but the metal must be present in order that the enzyme can function. Metals in this latter category are called co-enzymes. Sixteen metals have been established as co-enzymes, sodium, potassium, rubidium, caesium, magnesium, calcium, zinc, cadmium, copper, manganese, iron, cobalt, nickel, chromium, molybdenum, and aluminium. The presence of iron, molybdenum, and copper has also been demonstrated in prosthetic groups. Studies on the coordination compounds which metal ions form with organic molecules may lead to a deeper understanding of enzyme action.

Topic 17

Equilibrium and free energy

In this topic we shall be concerned mainly with establishing some important interconnections between some of the different parts of chemistry with which you are already familiar. In Topic 12 (Equilibria: gaseous and ionic) you have developed a method of describing the direction and extent of chemical change by means of equilibrium constants. In Topic 15 (Equilibria: redox and acid-base systems) you have applied this to redox and acid-base reactions. In Topic 7 you have seen how to forecast, on an approximate basis using enthalpies of formation, whether or not a given chemical compound is likely to be stable at room temperature. In Topic 12 you came across a hint as to the connection between standard enthalpy changes for reactions and the way in which equilibrium constants alter with temperature. In Topic 14 (Reaction rates) you have considered the kinetic aspects of the stabilities of substances. And in Topic 16 you have again tackled the problem of forecasting the direction and extent of chemical reactions, this time of ions in solution, in terms of standard redox potentials ($E^{\ominus}$ values).

In this topic all these ideas are drawn together and extended so that the links can be seen between

1 the various ways of describing the direction and extent of chemical reactions, and

2 equilibrium constants and energy changes.

17.1 The direction of a chemical change

The idea of the direction of a chemical change seems at first sight a simple one. Take one mole of red hot carbon atoms (graphite), and place it in a vessel containing one mole of oxygen molecules, O_2, and the reaction

$$C(\text{graphite}) + O_2(\text{gas}) \rightarrow CO_2(\text{gas})$$

takes place until all the carbon and oxygen have been converted into carbon dioxide. It is convenient to say that such a reaction 'goes to completion'. The direction of the change is 'in favour of the formation of carbon dioxide', etc. It is noticeable that the reaction is strongly exothermic. It is also useful to say that, in these conditions, the reverse process, that is, the conversion of carbon dioxide into carbon and oxygen, which is endothermic, 'does not take place'.

In many other systems however the situation is not so clear-cut. There are many cases where reactions do not 'go to completion' or 'fail to take place'. For example the conversion

$$N_2O_4(g) \rightarrow 2NO_2(g)$$

does not go to completion, nor does it fail to occur at ordinary temperatures. At 60 °C and 1 atm pressure, for example, about two thirds of the N_2O_4 has dissociated into NO_2 molecules; an equilibrium is set up for which K_p has a value of about 1.3 atm. Here it is more appropriate to speak of reactions occurring until a position of equilibrium is reached. Chemical reactions tend to equilibrium.

As was indicated in Topic 12, this idea that reactions tend to positions of equilibrium can be extended to include cases such as the combustion of hot carbon by saying that in these cases values of K are very large. Likewise for cases of 'non-reaction' K values are minutely small. It is possible, even in the case of burning carbon, that when the reaction is over there are traces of carbon and oxygen remaining, that is, K is not infinitely large, although the amounts may be so small as to be undetectable. In fact K has a calculated value of 10^{69}.

Kinetic barriers to chemical reaction

You will have noticed already, for example in Topic 7 (Energy changes and bonding), that there are many mixtures of chemical substances which can exist in a state far removed from equilibrium. At room temperature, for example, carbon and oxygen can coexist indefinitely without any observable conversion to carbon dioxide. 'Diamonds are forever.' We can put this another way by saying that the *rate* at which carbon and oxygen combine at room temperature is infinitesimal, although the value of the equilibrium constant for the carbon-oxygen-carbon dioxide system is 10^{69}.

The situation is different, however, when the carbon is red hot. In Topic 14 (Reaction rates) the idea of *activation energy* was discussed.

In order to do any converting of reactants to products, an energy barrier must be traversed.

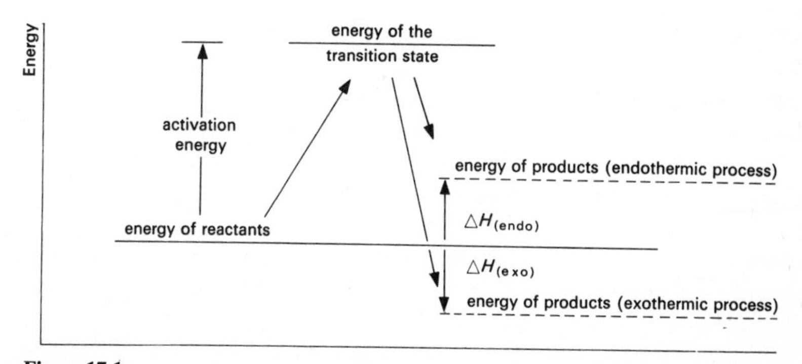

Figure 17.1
The path of minimum energy for a chemical reaction.

Figure 17.1 is an energy level diagram showing the energy changes involved in a chemical reaction. As you can see, there is a certain minimum energy called the activation energy which must be supplied to the reacting molecules before there is any possibility of reaction. At room temperature the proportion of the molecules of reactants having energies of comparable magnitude with that of the activation energy may be very small, so the rate at which the process proceeds may be very slow. At higher temperatures (e.g. at red heat) the proportion of molecules having energies of the same order as the activation energy is larger and the reaction is much faster. The necessity for the supply of the activation energy can therefore act as a 'barrier' preventing the attainment of equilibrium. Catalysts can also be used, to make available alternative routes from reactants to products for which the activation energy barriers are smaller.

Therefore, at high temperatures, or if catalysts can be found, the kinetic barrier to the attainment of equilibrium can be overcome and reactions can proceed to equilibrium relatively unimpeded. At room temperature, however, and in cases where catalysts are not available, this is not necessarily so, and mixtures of substances can coexist for indefinitely long periods in a state far removed from equilibrium. It is important to bear this in mind all the time in this topic. If it were not so, we should all have been converted to ammonia, carbon dioxide, and water vapour long ago.

17.2 How to find equilibrium constants for any particular (gas) reaction at different temperatures

Given that a chemical reaction is not kinetically hindered, it is often useful to know the extent to which it can be expected to take place under given conditions. For example, what will be the position of equilibrium, as measured by the equilibrium constant, in the 'water gas' reaction,

$$H_2O(g) + C(s) \rightleftharpoons H_2(g) + CO(g),$$

at 500 K; at 1000 K, or at 1500 K?

Clearly, it is not always a simple matter, sometimes not even possible, to answer these questions by direct experiment. It is therefore very useful to be able to *calculate* values of equilibrium constants from other data, and in this indirect way to obtain information from which to judge where the position of equilibrium will be, for example, for the water gas reaction at 1500 K. This section shows how to calculate the value of K_p for any gas reaction at various temperatures, if a value of K_p at another temperature is known.

You may remember that in Topic 12 (Equilibria: gaseous and ionic), when discussing the effect of temperature on the equilibrium constant for gas reactions, it was concluded that:

> an *increase in temperature* of an equilibrium system results in an *increase in the value of the equilibrium constant* if the reaction involved is *endothermic* ($\Delta H^{\ominus}$ positive), and a *decrease in the value of the equilibrium constant* if the reaction is *exothermic* ($\Delta H^{\ominus}$ negative).

That is, the effect of temperature on the value of the equilibrium constant depends on the sign of $\Delta H^{\ominus}$. In this section we shall take matters a stage further and study precisely how temperature, equilibrium constants, and $\Delta H^{\ominus}$ are related. This will enable us to calculate values for equilibrium constants for gas reactions at temperatures at which the equilibrium constants have not been directly determined. Table 1 in the appendix to this topic shows some values of equilibrium constants at various temperatures for reactions for which the standard enthalpy changes (at 298 K) are also quoted. Using the values, for example, for the reaction

$$N_2O_4(g) \rightleftharpoons 2NO_2(g),$$

try drawing a graph of K_p against temperature (in K).

Do you find any difficulties in accommodating the values of K_p on a linear scale within a reasonable compass?
Try plotting values of lg K against temperature. Is your graph a straight line? How would you describe its shape?

Now try plotting values of lg K_p against values of $\frac{1}{T}$. What do you infer about the relationship between lg K_p and $\frac{1}{T}$? Remembering that the general form of a straight line graph is

$$y = \mathrm{m}x + \mathrm{c}$$

where m is the gradient of the graph and c is the intercept on the y axis, what equation can you suggest for the graph of lg K_p against $\frac{1}{T}$? Notice that this is a reaction for which

$$\Delta H^{\ominus}_{298} \text{ is } +58.2 \text{ kJ.}$$

Repeat this work using, for example, the data quoted in appendix table 1 for the reaction

$$N_2(g) + 3H_2(g) \rightleftharpoons 2NH_3(g);\ \Delta H^{\ominus}_{298} = -92 \text{ kJ}$$

Do the results of your graphical work

1 agree with the general rule quoted from Topic 12 (Equilibria: gaseous and ionic) at the beginning of this section;

2 enable you to state, any more precisely than this rule does, the relationship between K_p, T, and $\Delta H^{\ominus}$?

You may have noticed that in both these cases the sign of the gradient of the graph of lg K_p against $\frac{1}{T}$ was opposite to the sign of $\Delta H^{\ominus}$. It can be shown in more advanced work that $\Delta H^{\ominus}$ enters into the relationship between lg K_p and $\frac{1}{T}$ as follows

$$\lg K_p \approx \text{constant} - \frac{\Delta H^{\ominus}}{2.3\,R}\left(\frac{1}{T}\right)^{*}$$

It follows from this that the gradient of the graph of lg K_p against $\frac{1}{T}$ should be

$$-\frac{\Delta H^{\ominus}}{2.3\,R}$$

From the gradients of your graphs estimate values of $\Delta H^{\ominus}$. How well do they agree with the quoted values of $\Delta H^{\ominus}_{298}$ for each reaction?

Does $\Delta H^{\ominus}$ vary with temperature?

The equation connecting lg K_p with $\frac{1}{T}$ was quoted as an approximate one because strictly speaking it would be an exact relation only if $\Delta H^{\ominus}$ did not vary at all with temperature. The fact that the graphs of lg K_p against $\frac{1}{T}$ do not depart markedly from straight lines even over quite big temperature ranges (five or six hundred degrees or more) is evidence that *in these temperature ranges* $\Delta H^{\ominus}$ does not change very much with temperature. It is for this reason that $\Delta H^{\ominus}_{298}$ can be used as an approximation for the $\Delta H^{\ominus}$ referred to in the above equation.

* The 2.3 arises in this expression because the actual relationship as first derived is

$$\ln K_p = \text{another constant} - \frac{\Delta H^{\ominus}}{R}\left(\frac{1}{T}\right)$$

where ln K_p means the logarithm of K_p to the base e and $\ln K_p = 2.3 \lg K_p$.

Some further information on the temperature-dependence of $\Delta H^{\ominus}$ can be obtained by looking at some of the data in table 2 of the appendix to this topic. Select by inspection the reaction for which $\Delta H^{\ominus}_T$ seems to vary most with temperature. Calculate the percentage change in $\Delta H^{\ominus}_T$ for a change of 100 kelvins in temperature.

Alternatively draw some graphs of $\Delta H^{\ominus}_T$ against T for various reactions. What do you notice about the gradients of these graphs in these temperature ranges?

Calculating values of K_p at various temperatures

Example 1

What would be the value of K_p for the equilibrium

$$2SO_2(g) + O_2(g) \rightleftharpoons 2SO_3(g)$$

at 800 K given the data in appendix table 1?

Probably the simplest way of tackling this would be to interpolate the value of $\lg K_p$ from a graph of $\lg K_p$ against $\frac{1}{T}$ and hence to calculate K_p.

Example 2

Suppose, however, that the data available were even more restricted than the selection of values in appendix table 1 and that the only values available were

$$2SO_2(g) + O_2(g) \rightleftharpoons 2SO_3(g);\ \Delta H^{\ominus}_{298} = -197 \text{ kJ}$$

and, at 500 K, K_p has a value 2.5×10^{10} atm^{-1}. Could we, using this information and the suggested relation

$$\lg K_p = \text{constant} - \frac{\Delta H^{\ominus}}{2.3\,R}\left(\frac{1}{T}\right),$$

find a value for K_p at 800 K?

This could be done as follows:

1 Substitute the available values of T, K_p, R, and $\Delta H^{\ominus}$ (that is, 500 K, 2.5×10^{10} atm^{-1}, 8.31 J K^{-1} mol^{-1}, and -197 kJ mol^{-1}) in the equation and so evaluate the constant.

2 Using the value of the constant so obtained, use the same expression again to calculate the value of K_p when $T = 800$ K. How well does this agree with the value you obtained from the graph?

For the more mathematically minded, this can be done for the general case as follows:
Suppose at a temperature T K, K_p has a value K_1 (to avoid confusion we shall omit the usual p subscript). Substituting as in Example 2, we get

$$\lg K_1 = \text{constant} - \frac{\Delta H^\ominus}{2.3\ RT_1}$$

$$\therefore \text{constant} = \lg K_1 + \frac{\Delta H^\ominus}{2.3\ RT_1}$$

At temperature T_2, let the equilibrium constant be K_2. Substituting again in the same formula and then putting in the value we have obtained for the constant, we get

$$\lg K_2 = \text{constant} - \frac{\Delta H^\ominus}{2.3\ RT_2}$$

$$\therefore \lg K_2 = \lg K_1 + \frac{\Delta H^\ominus}{2.3\ RT_1} - \frac{\Delta H^\ominus}{2.3\ RT_2}$$

$$\therefore \lg K_2 - \lg K_1 = \frac{\Delta H^\ominus}{2.3\ RT_1} - \frac{\Delta H^\ominus}{2.3\ RT_2}$$

$$\therefore \lg\left(\frac{K_2}{K_1}\right) = \frac{\Delta H^\ominus}{2.3\ R}\left(\frac{1}{T_1} - \frac{1}{T_2}\right)$$

17.3 How to find equilibrium constants (at a particular temperature) indirectly

In the previous section, we studied ways of calculating values of K_p at various temperatures. In this section we consider ways of determining equilibrium constants indirectly at one temperature, not from other K values, but from certain energy changes called standard free energy changes.

$\Delta G^\ominus$ and K

In section 17.2 you saw that $\Delta H^\ominus$, which is essentially an energy quantity, is related to the gradient of the graph of the logarithm of the equilibrium constant against the reciprocal of temperature for any particular reaction.
We now meet an energy quantity, called the standard free energy change for a chemical reaction, which is directly related to the logarithm of the equilibrium constant as follows:

$$\Delta G^\ominus = -2.3\ RT \lg K^*$$

* The 2.3 enters this equation because the relation between $\Delta G^\ominus$ and K, as derived from advanced theory, is $\Delta G^\ominus = -RT \ln K$.

$\Delta G^{\ominus}$ is sometimes called the Gibbs standard free energy change after the American physical chemist J. Willard Gibbs.

Two things are at once apparent from this relationship.

1 Since RT is an energy quantity (expressed for example in kJ mol^{-1}) and lg K is a pure number, $\Delta G^{\ominus}$ is an energy quantity.

2 $\Delta G^{\ominus}$ is a temperature-dependent quantity since T is in this equation as well as lg K. $\Delta G^{\ominus}$ values must therefore be quoted for some particular temperature. A useful standard temperature at which many values are quoted is 298 K (25 °C) and standard free energy changes at this temperature are written $\Delta G^{\ominus}_{298}$.

Figure 17.3a shows the way in which $\Delta G^{\ominus}$ and lg K are related at 298 K.

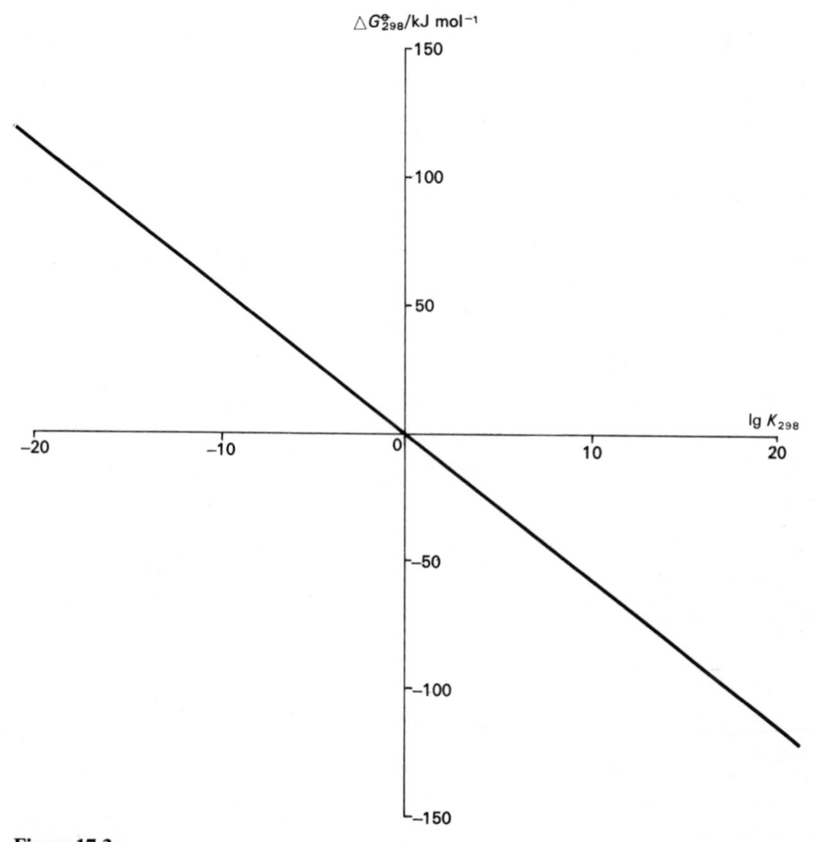

Figure 17.3a
Graph of $\Delta G^{\ominus}_{298}$ plotted against lg K_{298}. (Since $\Delta G^{\ominus}_{298} = -2.303\ RT$ lg K, the gradient of this graph is $-2.303\ RT$. At 298 K, as here, its value is -5.69 kJ mol^{-1}.)

You may wonder what the graph of $\Delta G^{\ominus}_{298}$ plotted against K itself would look like and wish to draw such a graph for yourself. You will need a long strip of graph paper in order to do so!

There are several points to notice about figure 17.3a.

1 For values of K greater than about 10^{10}, $\Delta G^{\ominus}_{298}$ is negative and has values less than -60 kJ mol^{-1}. This corresponds to a situation in which the reaction has gone 'virtually to completion' (see section 17.1).

2 For very small values of K (less than about 10^{-10}) $\Delta G^{\ominus}_{298}$ is positive and has values greater than $+60$ kJ mol^{-1}. This corresponds to a situation in which the reaction has 'failed to occur'.

3 In the intermediate range (10^{-10} to 10^{10}) of values of K, which includes all those we usually refer to as equilibrium reactions, $\Delta G^{\ominus}_{298}$ has values in the range $+60$ kJ mol^{-1} to -60 kJ mol^{-1}. For equilibria in which the position of equilibrium is 'over on the righthand side' (that is, when products predominate in the equilibrium mixture) $\Delta G^{\ominus}_{298}$ has small negative values. For equilibria in which the position of equilibrium is 'on the lefthand side' (reactants predominate) $\Delta G^{\ominus}_{298}$ has small positive values. We may summarize all this as shown in table 17.3a.

Standard free energy change $\Delta G^{\ominus}_{298}$/kJ mol^{-1}	**Equilibrium constant K_{298}**	**Extent of reaction**
less than -60	10^{10} and larger	reaction 'complete'
between -60 and 0	10^{10} to 1	products predominate in an 'equilibrium'
between 0 and $+60$	1 to 10^{-10}	reactants predominate in an 'equilibrium'
greater than $+60$	10^{-10} and smaller	'no' reaction

Table 17.3a
The relationship between standard free energy and equilibrium constant for a reaction

Questions

1 K_p for the equilibrium

$$N_2(g) + 3H_2(g) \rightleftharpoons 2NH_3(g)$$

is 6.76×10^5 atm^{-2} at 298 K.
What is $\Delta G^{\ominus}_{298}$?

2 For the reaction

$$ZnO(s) + CO(g) \rightleftharpoons Zn(s) + CO_2(g)$$

the ratio $\frac{p_{CO_2 eqm}}{p_{CO\ eqm}}$ at 298 K is 2×10^{-11}.
What is $\Delta G^{\ominus}_{298}$ for this reaction?

3 For the Daniell cell reaction

$$Cu^{2+}(aq) + Zn(s) \rightleftharpoons Cu(s) + Zn^{2+}(aq)$$

$\Delta G^{\ominus}_{298}$ is -212 kJ.
What is K_c?

$\Delta G^{\ominus}$ and $\Delta H^{\ominus}$

You may recall that in Topic 7 (Energy changes and bonding) we tacitly assumed that if a chemical reaction was strongly exothermic (ΔH negative) then the products of that reaction would be 'energetically stable', that is K for the reaction would be large, and, correspondingly, if a chemical reaction was strongly endothermic (ΔH positive) then the products would be 'energetically unstable', that is, K would be small. You may also remember that this assumption was hedged about with several provisos of which the following were most important:

1 The ΔH values (positive or negative) must be considerable in magnitude, for example more than 40 or 50 kJ mol^{-1}.

2 The temperature must not be too far removed from room temperature (that is 298 K).

3 There are exceptions to this rule; some endothermic chemical processes are known which occur spontaneously even at ordinary temperatures.

You can now see that the appropriate energy quantity which is related to the value of the equilibrium constant in this way (that is large and positive for a reaction which does not occur and large and negative for one which goes to completion) is not $\Delta H^{\ominus}$ but $\Delta G^{\ominus}$. The degree to which the assumption made in Topic 7 is justified therefore depends on how closely $\Delta H^{\ominus}$ and $\Delta G^{\ominus}$ values agree.

Look at the values of $\Delta H^{\ominus}_{298}$, $\Delta G^{\ominus}_{298}$, and K_{298} for various chemical reactions which are given in table 3 in the appendix to this topic.

1 When K is very large, are $\Delta H^{\ominus}_{298}$ and $\Delta G^{\ominus}_{298}$ of the same sign? Are they of the same order of magnitude?

2 When K is very small, are $\Delta H^{\ominus}_{298}$ and $\Delta G^{\ominus}_{298}$ of the same sign? Are they of the same order of magnitude?

3 When K has intermediate values (between 10^{-10} and 10^{10}), is there good agreement between $\Delta H^{\ominus}_{298}$ and $\Delta G^{\ominus}_{298}$?
What do you notice about the types of reaction for which agreement is not good?

Look at the corresponding data in table 4 of the appendix and answer the above questions (where appropriate) using these data.

Now look in table 2 at values of $\Delta H^{\ominus}$ and $\Delta G^{\ominus}$ at various temperatures for a selection of reactions involving gases.

1 When $T = 298$ K, do $\Delta H^{\ominus}$ and $\Delta G^{\ominus}$ ever differ in sign in this set of examples?

2 At higher temperatures is the difference between $\Delta H^{\ominus}$ and $\Delta G^{\ominus}$ greater or less than at room temperature?

3 Is the value of $\Delta H^{\ominus}$ generally a good indication of the position of equilibrium at high temperatures?

This section (17.3) is entitled 'How to find equilibrium constants (at a particular temperature) indirectly'. You may therefore be wondering why we have entered into a discussion of the standard free energy change, $\Delta G^{\ominus}$. Clearly, however, if there are other ways of determining values of $\Delta G^{\ominus}$ (other, that is, than by measuring equilibrium constants) then the values of $\Delta G^{\ominus}$ so obtained can be used to calculate equilibrium constants for reactions for which the direct measurement of K is difficult or impossible.

We next, however, consider another way of determining equilibrium constants, namely from $E^{\ominus}$ values, since this will incidentally, also throw some light for us on the nature of standard free energy changes.

Using $E^{\ominus}$ values to forecast the direction of chemical change

For reactions taking place in solution you have already come across, in Topic 16, a way of predicting the likely direction of a chemical change, and therefore the likely magnitude of the equilibrium constant, using $E^{\ominus}$ values. You will find it useful to revise section 16.1 before continuing.

Revision question

Use the standard redox potentials listed in the *Book of Data* to predict whether:

1 Bromine water will oxidize Fe^{2+}(aq) ions to Fe^{3+}(aq).

2 Bromine water will oxidize:

a I^{-}(aq) ions to I_2(aq)

b Cl^{-}(aq) ions to Cl_2(aq).

3 Sn^{2+}(aq) ions will reduce Fe^{3+}(aq) ions.

4 Zinc dust will reduce $Fe^{3+}(aq)$ ions.
5 Acidified potassium permanganate solution will oxidize $Br^{-}(aq)$ ions.
6 Acidified potassium permanganate will oxidize copper.

You may like to test your predictions by means of the following set of simple test-tube experiments.

Experiment 17.3a

To test predictions about some redox reactions

1 Add bromine water to iron(II) ammonium sulphate solution. Test for iron(III) ions before and after addition and note any colour change during reaction.

2 Add bromine water to:
a potassium iodide solution
b potassium chloride solution.

3 Add tin(II) chloride solution (Sn^{2+} ions) to a solution containing iron(III) ions [iron(III) alum, iron(III) sulphate, or iron(III) chloride]. Test for iron(II) ions before and after addition.

4 Add zinc dust to a solution containing iron(III) ions. Cork the test-tube and shake well. Test for formation of iron(II) ions.

5 Add acidified potassium permanganate solution drop by drop to potassium bromide solution. Note the colour changes and shake the mixture with a few drops of tetrachloromethane.

6 Add a *little dilute* potassium permanganate solution to a suspension of copper powder in dilute sulphuric acid. Cork the tube and shake for 2–3 minutes. Can you detect any signs of reaction? If in doubt, decant some of the final solution into another test-tube and add ammonia solution (approximately 2M).

Questions

1 Were your predictions confirmed by experiment?
2 How did you use the $E^{\ominus}$ values to make these predictions?

Your answer to this last question (question 2) will probably have been in terms of the rule formulated in section 16.1 on the d-block elements. A slightly different way of formulating the same rule, and one which will be more useful for our purposes in this section is given in the following example.

Consider the problem of predicting whether zinc dust will reduce $Fe^{3+}(aq)$ ions. The proposed process is

$$Zn(s) + 2Fe^{3+}(aq) \rightarrow Zn^{2+}(aq) + 2Fe^{2+}(aq)$$

for which the cell diagram is

$$Zn(s) \mid Zn^{2+}(aq) \vdots Fe^{3+}(aq), Fe^{2+}(aq) \mid Pt$$

Applying the rules for finding the standard e.m.f.s of cells (see Topic 15, Equilibria: redox and acid-base systems), we find that $E^\ominus_{cell} = -(-0.76) + 0.77 = +1.53$ V. Since $E^\ominus_{cell}$ is positive in sign and has a magnitude greater than about 0.4 V, we can safely predict that the cell reaction is likely to take place in the direction indicated by the equation. We can summarize this rule as in table 17.3b.

$E^\ominus_{cell}$	K_c for cell reaction
+0.4 V or larger	very large
−0.4 V or less	very small

Table 17.3b
Cell e.m.f.s and K_c for cell reactions

A glance at the values of $E^\ominus_{cell}$ and $K_{c,298}$ in appendix table 4 shows that this rule is a reasonable working guide to the direction of chemical reactions.

More precise relationship between $E^\ominus_{cell}$ and K_c

In order to be able to calculate equilibrium constants of cell reactions, we need to know more precisely how $E^\ominus_{cell}$ and K_c are related. This we can see using the Nernst equation.

You will recall (Topic 15, Equilibria: redox and acid-base systems) that for a metal/ion electrode system the electrode potential, relative to the standard hydrogen electrode, is given by a simple form of the Nernst equation as

$$E = E^\ominus + \frac{2.3\,RT}{zF} \lg\,[\text{ion}]$$

for a particular temperature (usually 298 K). [ion] refers to the *actual* concentration of the ion present in the half-cell (not necessarily an equilibrium concentration) and $E^\ominus$ is the standard redox potential of the electrode.

Example
Nernst equation for the Daniell cell

Figure 17.3b shows a diagram of a typical Daniell cell. Inside the porous pot is the zinc half-cell which has an electrode potential relative to hydrogen given by the Nernst equation as

$$E_{Zn} = E^{\ominus}_{Zn} + \frac{2.3\,RT}{zF}\lg\,[Zn^{2+}(aq)]$$

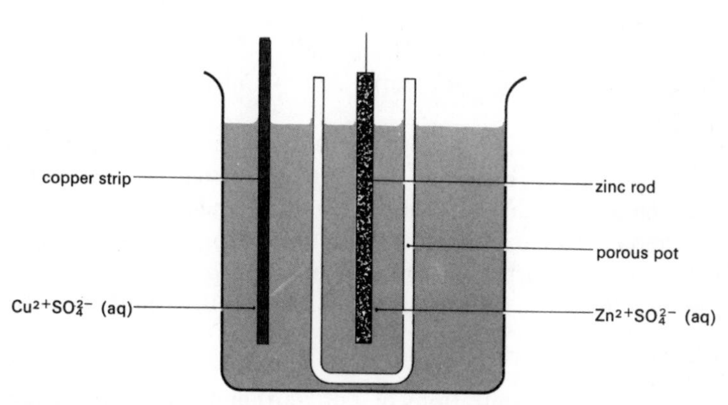

Figure 17.3b
Diagram of a Daniell cell. The concentrations of $Zn^{2+}(aq)$ and $Cu^{2+}(aq)$ are $[Zn^{2+}(aq)]$ and $[Cu^{2+}(aq)]$ and not necessarily 1M.

Similarly the electrode potential of the copper electrode relative to hydrogen is

$$E_{Cu} = E^{\ominus}_{Cu} + \frac{2.3\,RT}{zF}\lg\,[Cu^{2+}(aq)]$$

The e.m.f. of this Daniell cell is found by taking the difference between E_{Cu} and E_{Zn}:

$$E_{cell} = E_{Cu} - E_{Zn}$$

$$= E^{\ominus}_{Cu} - E^{\ominus}_{Zn} \quad - \left(\frac{2.3\,RT}{zF}\lg\,[Zn^{2+}(aq)] - \frac{2.3\,RT}{zF}\lg\,[Cu^{2+}(aq)]\right)$$

$$\therefore\ E_{cell} = E^{\ominus}_{cell} \quad - \frac{2.3\,RT}{zF}\lg\frac{[Zn^{2+}(aq)]}{[Cu^{2+}(aq)]}$$

This equation is a simple form of the Nernst equation for the Daniell cell. $E^{\ominus}_{cell}$ is the standard e.m.f. of the cell and has a value of 1.1 V.

Short-circuiting the cell

Suppose now that this cell is short-circuited by connecting the zinc pole to the copper pole by means of a conducting wire. What would happen?

When the cell is short-circuited, in the external circuit, electrons pass from the zinc pole along the wire to the copper pole. Inside the cell the reaction

$$Zn(s) + Cu^{2+}(aq) \rightarrow Zn^{2+}(aq) + Cu(s)$$

takes place, that is,
at the zinc electrode: $Zn(s) \rightarrow Zn^{2+}(aq) + 2e^-$
and at the copper electrode: $Cu^{2+}(aq) + 2e^- \rightarrow Cu(s)$

These processes all continue as long as there remains a potential difference between the two half-cells to drive the electrons through the connecting wire, that is, as long as E_{cell} has some positive value. When there is no longer any potential difference, there is no net transfer of electrons along the wire. Correspondingly, there is no further net transfer of electrons in the solution between the zinc and copper. The cell reaction has attained the equilibrium situation:

$$Zn(s) + Cu^{2+}(aq) \rightleftharpoons Zn^{2+}(aq) + Cu(s)$$

In this situation the concentrations of $Zn^{2+}(aq)$ and $Cu^{2+}(aq)$ are the equilibrium concentrations, $[Zn^{2+}(aq)]_{eqm}$ and $[Cu^{2+}(aq)]_{eqm}$, and K_c for this equilibrium is $\dfrac{[Zn^{2+}(aq)]_{eqm}}{[Cu^{2+}(aq)]_{eqm}}$.

(In fact K_c for this particular equilibrium has a value of about 10^{37}, so that in practical terms the reaction has gone to completion, and either all the zinc or all the copper sulphate solution has been 'used up'.)

Applying the Nernst equation for the Daniell cell at equilibrium, we put in

$$E_{cell} = 0$$

and $[Zn^{2+}(aq)] = [Zn^{2+}(aq)]_{eqm}$
and $[Cu^{2+}(aq)] = [Cu^{2+}(aq)]_{eqm}$

$$\therefore 0 = E^{\ominus}_{cell} - \frac{2.3\,RT}{zF} \lg \frac{[Zn^{2+}(aq)]_{eqm}}{[Cu^{2+}(aq)]_{eqm}}$$

$$\therefore E^{\ominus}_{cell} = \frac{2.3\,RT}{zF} \lg K_c.$$

In this expression the units of R are $J\,K^{-1}\,mol^{-1}$, that is $R = 8.31\,J\,K^{-1}\,mol^{-1}$
With $F = 96\,500$ C, this gives $E^{\ominus}$ in volts.

To summarize, then, when a cell is short circuited, its e.m.f. falls to zero, the cell reaction attains equilibrium, and the Nernst equation reduces to an expression connecting $E^{\ominus}_{cell}$ and K_c for the cell equilibrium. We can therefore calculate values of K_c for cell reactions indirectly from $E^{\ominus}$ values. In fact, in using $E^{\ominus}$ values predictively we have merely been using equilibrium constants in a sort of Nernst disguise.

Table 4 in the appendix to this topic lists some equilibrium constants for a selection of cell reactions. These K_c values have been calculated from the corresponding $E^{\ominus}_{cell}$ values using the Nernst equation.

Questions

The cell

$$Cu(s) \mid Cu^{2+}(aq) \,\vdots\, Br_2(aq), 2Br^-(aq) \mid Pt$$

is set up and short-circuited.

1 Write an equation for the resulting cell equilibrium reaction.
2 Calculate K_c for this equilibrium.

(Standard redox potentials are given in the *Book of Data*.)

Experiments 17.3b and 17.3c

Comparison of values of an equilibrium constant obtained indirectly and directly

The aim of these next two experiments is to enable you to obtain values of K_c for a chemical reaction

1 indirectly, by measurements of cell e.m.f.s (experiment 17.3b), and
2 directly by titration (experiment 17.3c)

and to compare these values to see how closely they agree.

Experiment 17.3b

Finding the equilibrium constant for the reaction $Ag^+(aq) + Fe^{2+}(aq) \rightleftharpoons Fe^{3+}(aq) + Ag(s)$ by e.m.f. measurements on the cell $Pt \mid Fe^{2+}(aq), Fe^{3+}(aq) \,\vdots\, Ag^+(aq) \mid Ag(s)$

You will require

0.10M iron(III) nitrate in 0.05M nitric acid (approximately 5 cm^3 of concentrated HNO_3 per cubic decimetre of solution)
0.20M iron(II) sulphate
0.20M barium nitrate
0.40M silver nitrate
ammonium nitrate/gelatine salt bridge
platinum electrode (foil or wire)
silver electrode (foil or wire)
pH meter or potentiometer reading to 1 millivolt

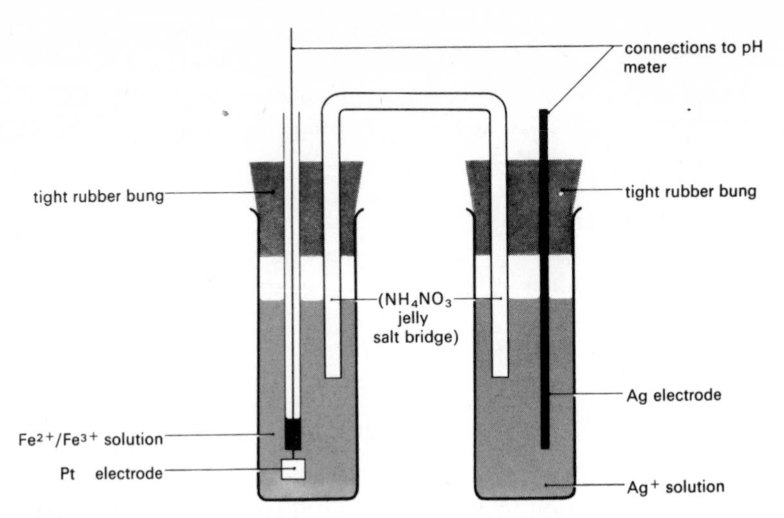

Figure 17.3c

Preparation of the Fe^{2+}(aq)/Fe^{3+}(aq) half-cell

This half-cell may be supplied to you ready for use, but in case you need to make one up for yourself the instructions are as follows.

Mix equal volumes of 0.20M $FeSO_4$ and 0.20M $Ba(NO_3)_2$ and allow the precipitate to settle for a few minutes in a stoppered vessel. The supernatant solution* is 0.10M with respect to iron(II) nitrate. Mix equal volumes of the 0.1M iron(II) nitrate and iron(III) nitrate solutions and transfer the mixture into the Fe^{2+}(aq)/Fe^{3+}(aq) half-cell. Seal this half-cell with a tightly fitting rubber bung fitted with the platinum electrode and salt bridge to avoid any oxidation of the Fe^{2+}(aq) ions by the air.

Procedure

Use the 0.40M silver nitrate solution to set up the silver half-cell as shown in the diagram. Measure and record the e.m.f. and polarity of the electrodes. Repeat the procedure with silver nitrate solutions of molarities 0.20, 0.10, 0.050, 0.025, prepared by successive dilution of the original 0.40M solution.

If you are measuring the e.m.f.s with a pH meter you will find it necessary to change over the connectors to the cell at some stage in the measuring series.

* Small quantities of barium sulphate may remain suspended in the solution and will not affect the results adversely.

Results

$[Ag^+(aq)]$ /mol dm^{-3}	lg $[Ag^+(aq)]$	$-$lg $[Ag^+(aq)]$	E_{cell} /V	Polarity of Ag half-cell	Polarity of $Fe^{2+}(aq)/Fe^{3+}(aq)$ half-cell

Plot the measured e.m.f. values against $-$lg $[Ag^+(aq)]$ and draw on the graph the straight line that best fits the plotted data. At what point on this straight line does $[Ag^+(aq)]$ have its equilibrium value, $[Ag^+(aq)]_{eqm}$?

Find from your graph the equilibrium concentration of $Ag^+(aq)$ for the cell reaction $Ag^+(aq)+Fe^{2+}(aq) \rightleftharpoons Fe^{3+}(aq)+Ag(s)$ and calculate

$$K_c = \frac{[Fe^{3+}(aq)]_{eqm}}{[Ag^+(aq)]_{eqm}[Fe^{2+}(aq)]_{eqm}}$$

Note

An alternative to the above procedure would be to alter the Fe(II)/Fe(III) ratio by a continuous variation method, whilst maintaining a fixed silver ion concentration in the other half-cell. This leads to an identical result for K, but the e.m.f. values observed are smaller than those measured in the suggested procedure, thus causing larger experimental errors.

Experiment 17.3c

To find the equilibrium constant for the reaction $Ag^+(aq) + Fe^{2+}(aq) \rightleftharpoons Fe^{3+}(aq) + Ag(s)$ by titration

You will require

0.10M iron(II) nitrate, prepared as described in experiment 17.3b
0.10M silver nitrate
and *either*
1M sodium chloride
phosphoric acid
standardized potassium permanganate solution, approximately 0.02M
or
standardized potassium thiocyanate solution, approximately 0.1M

Procedure

Prepare several samples of reaction mixture by adding together 25.0 cm^3 each of the iron(II) nitrate and silver nitrate solutions in several dry 100 cm^3 conical flasks. Stopper the flasks tightly (why?) and leave them to stand for about half an hour, after which time it may be assumed that equilibrium has been attained.

Carefully pipette out into a conical flask 25.0 cm^3 of the supernatant solution above the silver precipitate and titrate the silver ion present by one of the following methods.

Method 1

Add to the mixture in the order given,

5 cm^3 1M NaCl
10 cm^3 dilute H_2SO_4
and 1–2 cm^3 concentrated H_3PO_4,

and then titrate the mixture with standardized potassium permanganate solution.

The chemical reactions upon which this titration depends are as follows:
On addition of the salt solution

$$Ag^+(aq) + Cl^-(aq) \rightarrow AgCl(s)$$

On addition of the potassium permanganate

$$MnO_4^-(aq) + 8H^+(aq) + 5e^- \rightarrow Mn^{2+}(aq) + 4H_2O(l)$$
$$Fe^{2+}(aq) \rightarrow Fe^{3+}(aq) + e^-,$$

so that one cubic decimetre of 0.02M potassium permanganate combines with

$\frac{1}{50} \times 5 = \frac{1}{10}$ mole of electrons, or one cubic decimetre of 0.1M $Fe(NO_3)_2$, etc.

Method 2

Titrate the 25 cm^3 sample of solution with standardized potassium thiocyanate solution. The chemical reactions taking place are

1 Reaction between $Ag^{2+}(aq)$ and $CNS^-(aq)$:

$$Ag^+(aq) + CNS^-(aq) \rightarrow AgCNS(s).$$

2 When all $Ag^+(aq)$ ions have been removed, thiocyanate ions react with the iron(III) ions also present in the equilibrium mixture to give a blood-red colour which acts as an indicator.

$$Fe^{3+}(aq) + CNS^-(aq) \rightarrow \underset{\text{blood red colour}}{FeCNS^{2+}(aq)}$$

Calculation of K_c

1 How many moles of Fe^{2+}(aq) (Method 1) or Ag^{+}(aq) (Method 2) did you find in the 25 cm^3 sample of the equilibrium mixture?

2 Therefore what was the concentration of Fe^{2+}(aq), or Ag^{+}(aq) in mol dm^{-3} of the equilibrium mixture?

3 When the equilibrium is set up from the starting solutions, how many moles of Fe^{3+}(aq) are formed for every mole of silver atoms precipitated?

4 Therefore, in what proportion will $[Ag^{+}(aq)]_{eqm}$ and $[Fe^{2+}(aq)]_{eqm}$ be in the equilibrium solution?

5 Bearing in mind that the initial concentrations of Ag^{+}(aq) and Fe^{2+}(aq) are 0.05M per cubic decimetre of equilibrium mixture, calculate $[Fe^{3+}(aq)]_{eqm}$.

6 Calculate K_c.

Comparison of values of K_c obtained by the two experiments

1 How well (within how many per cent) do your two values (one from experiment 17.3b and the other from experiment 17.3c) agree? How can you account for any discrepancies between them?
Calculate the value of K_c using the $E^\ominus$ values (*Book of Data*) for the half-cells used in experiment 17.3b.

2 Are your experimental values of K_c (*a*) higher or (*b*) lower than those calculated from $E^\ominus$ values? Can you suggest any reasons for this?

$-\Delta G^\ominus$ as the work of a cell reaction

Earlier in this section $\Delta G^\ominus$ was defined as

$$\Delta G^\ominus = -2.3\ RT \lg K$$

(for some particular temperature)

From the Nernst equation for cell reactions

$$E^\ominus = +\frac{2.3\ RT}{zF} \lg K_c$$

(at a particular temperature)

Comparing these two equations,

$$\Delta G^\ominus = -zFE^\ominus$$

$\Delta G^\ominus$ is an energy and $zFE^\ominus$ is a work quantity (in units of joules)*

* It may be useful to recall, purely as a memory aid, that
power ($J\ s^{-1}$) = p.d.(V) × current (A)
i.e. rate of doing work ($J\ s^{-1}$) = p.d.(V) × rate of transfer of charge ($C\ s^{-1}$)
∴ work (J) = p.d.(V) × charge(C)

The equation $\Delta G^{\ominus} = -zFE^{\ominus}$ therefore enables us to picture the standard free energy change for a cell reaction as the work a cell would be expected to do if the potential difference across its terminals were its e.m.f. Such work might be the driving of an electric motor to raise a weight, etc. At first sight this seems a paradox since we define the e.m.f. of a cell as that p.d. across the terminals of the cell when the cell is on open circuit (that is, doing no work). We can nevertheless calculate a value for this limiting available work. We can also find by a simple experiment (experiment 17.3d) the limit to which the work obtainable from a cell tends as the resistance between its terminals is increased.

Experiment 17.3d
The work of a cell reaction

You will require

a Daniell cell with amalgamated zinc rod
molar solutions of copper(II) sulphate and zinc sulphate
high resistance voltmeter (0–3 V) or potentiometer
milliammeter
connecting leads
rheostat (up to 1 or 2 kΩ)
graph paper

Cell diagram

$Zn(s) \mid Zn^{2+}(aq) \,\vdots\, Cu^{2+}(aq) \mid Cu(s)$

Alternatively you may be asked to study one of the following cells:

$Zn(s) \mid Zn^{2+}(aq) \,\vdots\, Ag^{+}(aq) \mid Ag(s)$
$Cu(s) \mid Cu^{2+}(aq) \,\vdots\, Ag^{+}(aq) \mid Ag(s)$

Circuit diagram

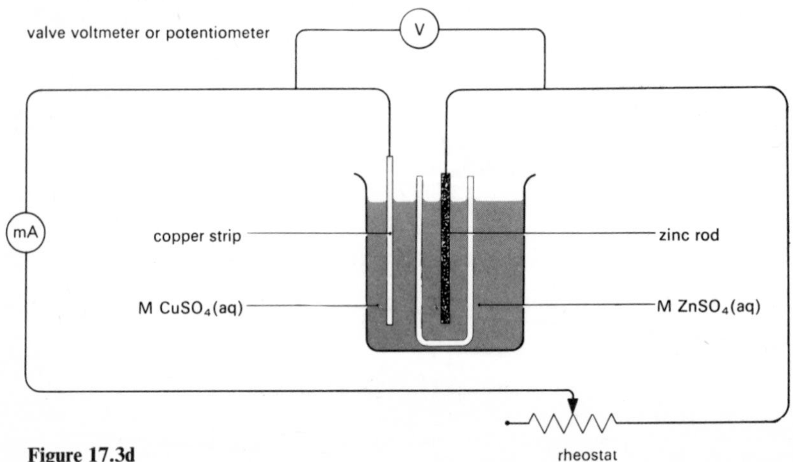

Figure 17.3d

Procedure

Set the rheostat to its lowest setting, and measure the cell voltage and current. Note your readings in a table.

p.d. across cell terminals/V	Current/mA

Increase the resistance until the p.d. is about 0.1 V higher, but do not waste time trying to make the interval exactly 0.1 V. Note the new voltage and current in your table. Repeat the procedure until the whole resistance of the rheostat is in circuit.

Treatment and discussion of results

Plot a graph of current against the cell p.d. (figure 17.3e).

Figure 17.3e

Questions

1 Is the form of the graph as you expected?

Extrapolate your graph to zero current.

2 What is the value of the p.d. at 'zero current' called?

3 What is its value from your graph?

4 Calculate the work per mole of the cell reaction which the cell could do at each of the p.d. values recorded.

a Write the overall ionic equation for the cell reaction.

b When one mole of zinc displaces one mole of copper ions what is the total charge in coulombs that is transferred? (If z electrons are involved on each side of the overall ionic equation the total charge transferred is zF coulombs where F is the Faraday.)

c For a cell voltage V the work of the reaction is therefore zFV J.

Note the results in a table.

Cell p.d./V	Work obtainable per mole of reaction/J

Plot a graph of the work obtainable per mole of reaction against the cell potential difference.

5 What is the limiting (maximum) work that could ever be obtained from the cell? To what value of the cell p.d. does this limiting work value correspond?

6 Compare the limiting work of the cell with the standard free energy change for the Daniell cell reaction quoted in table 4 in the appendix to this topic.

Time for reaction

Another way of thinking about the conditions under which the work of the cell may be made to approach its limiting value is to consider the way in which the time it takes for one g-equation of reaction to take place varies with cell voltage, as follows.

Current i is measured in units of coulombs per second (amperes), so if $t_{(reaction)}$ is the time for one mole of reaction,

$$i = \frac{zF}{t_{(reaction)}} \quad \text{coulombs per second}$$

$$\therefore t_{(reaction)} = \frac{zF}{i} \text{ seconds.}$$

As the current tends to zero, V approaches the e.m.f. value, so that, theoretically, infinite time would be required for the cell to produce the limiting available work. This may be seen from a graph of $t_{(reaction)}$ against the p.d. of the cell (figure 17.3f).

Figure 17.3f

You should now plot this graph.

Under all conditions where the cell is on load, V, the p.d. across the cell terminals is less than E, the cell e.m.f., so that the work obtainable, zFV, is less than zFE, that is, it lies between zero (when the cell is short circuited) and $-\Delta G$ (when the cell is on 'open circuit'). In practice the work which can be done by a cell in a circuit of high resistance can be made to approach the $-\Delta G$ value as closely as we wish, but the closer we wish to approach it the longer it will take. Fortunately we can obtain the cell e.m.f. value when the cell is practically on 'open circuit' and so we can calculate the limiting work the cell could be made to do without waiting for it actually to do any.

This interpretation of $\Delta G^\ominus$ as – (the limiting value of the work obtainable from a cell) enables us to think of $\Delta G^\ominus$ as a *difference* (equal to $-zFE$) between energies which we may associate with the products and reactants of the reaction. Such energies are called the standard free energies, $G^\ominus$, of the products and reactants

i.e. $\Delta G^\ominus = G^\ominus \text{(products)} - G^\ominus \text{(reactants)}$

The individual free energies cannot themselves be measured, but only the energy differences, for example as $-zFE^\ominus$. We shall take this point further in the next section (17.4).

Other ways of finding $\Delta G^\ominus$

So far we have seen how standard free energy changes can be determined for reactions between ions in solution. There are also other methods by which $\Delta G^\ominus$ can be determined for other reactions such as those between gases. Such methods are independent of the direct measurement of equilibrium constants for these reactions, so that they afford a further, indirect, way of determining the equilibrium constants. One method, for example, involves calorimetry at room temperatures and at temperatures down towards the absolute zero of temperature. The details of these methods, however, involve work more advanced than that which we shall undertake here.

17.4 Free energy in chemistry

The main aims of this section are

1 to enable you to use standard free energy data, as published in tables, to calculate equilibrium constants for chemical reactions, and

2 to show how free energy data can be interpreted in terms of the relative energetic stabilities of compounds.

The standard free energy of formation, $\Delta G_f^\ominus$

The definitions and conventions introduced here are exactly parallel to those concerning the standard heat of formation, $\Delta H_f^\ominus$. You may find it useful to revise section 7.1 (Energy changes and bonding) before continuing with this.

At the end of section 17.3 it was suggested that $\Delta G^\ominus$ be regarded as an energy *difference* between free energy values which we can assign to the reactants and products of a reaction. For example in the reaction

$$C(s) + O_2(g) \rightarrow CO_2(g)$$

we can say that

$$\Delta G_{298}^\ominus = G_{298}^\ominus[CO_2(g)] - \left\{G_{298}^\ominus[C(s)] + G_{298}^\ominus[O_2(g)]\right\}$$

where $G_{298}^\ominus$ [X] means 'the standard free energy of X'.

Just as in the case of enthalpies, or of potential energy, etc., absolute values of free energies are not known, so it is convenient to choose some base line, or arbitrary zero, from which to measure the standard free energies of substances. The convention chosen for free energies is the same as that for enthalpies, namely that

at 760 mmHg pressure
298 K
with the elements in the physical states normal under these conditions

the standard free energies of the elements are zero.

It should be emphasized that this convention is quite arbitrary, as arbitrary as calling the potential energy of all objects at sea level zero, but it is an agreed convention.

It follows necessarily from this convention that

$$\Delta G^{\ominus}_{f,298}\ [\text{element in physical state normal at 760 mmHg and 298 K}] = 0.$$

It is possible, therefore, using this convention, to tabulate standard free energies of formation of *compounds* rather than standard free energies relating to specific reactions. In fact, of course, the standard free energy of a compound does really refer to a reaction, namely the formation of one mole of the compound from its elements in physical states normal at 760 mmHg and 298 K. For example,

$$C(s) + O_2(g) \rightarrow CO_2(g);\ \Delta G^{\ominus}_{298} = -395\ \text{kJ}$$

and $\Delta G^{\ominus}_{f,298}\ [CO_2(g)] = -395\ \text{kJ}$

are exactly equivalent statements.

The tabulation of standard free energy data relating to compounds rather than reactions, however, makes for very general and flexible use of the tables as the examples later in this section will show.

How to calculate equilibrium constants from standard free energies of formation

As we have already seen (section 17.3), given the value of $\Delta G^{\ominus}$ for any chemical reaction, it is a simple matter to calculate the equilibrium constant for that reaction. The following examples show how K values can be calculated for a selection of reactions, given that values of $\Delta G^{\ominus}_f$ are available from tables.

1 *The equilibrium constants of some gas phase reactions*

a $N_2(g) + 3H_2(g) \rightleftharpoons 2NH_3(g)$

$$\Delta G^{\ominus}_{298} = 2\Delta G^{\ominus}_{f,298}\ [NH_3(g)] - \Delta G^{\ominus}_{f,298}\ [N_2(g)] - 3\Delta G^{\ominus}_{f,298}\ [H_2(g)]$$

Since the standard free energies of formation of elements for their standard states are zero (by convention)

$$\begin{aligned}\Delta G^{\ominus}_{298} &= 2\times\Delta G^{\ominus}_{f,298}\,[NH_3(g)]\\ &= 2\times(-17)\,\text{kJ (from tables, for example table 6 in the appendix to this topic)}\\ &= -34\,\text{kJ}\end{aligned}$$

By definition

$$\begin{aligned}\Delta G^{\ominus} &= -2.3\,RT\lg K_p\\ \therefore \Delta G^{\ominus}_{298} &= -2.3\times 8.3\times 10^{-3}\times 298\times \lg K_p\\ \therefore -34 &= -5.69\lg K_p\\ \therefore \lg K_p &= 5.98\end{aligned}$$

$$\therefore K_p = \frac{p^2_{NH_3eqm}}{p_{N_2eqm}\times p^3_{H_2eqm}} = 9.6\times 10^5\,\text{atm}^{-2}\text{ at 298 K}$$

thus illustrating that, if a catalyst could be found, and equilibrium could be reached, equilibrium mixtures of hydrogen and nitrogen at room temperature and pressure would be expected to contain mainly ammonia and only small amounts of unreacted hydrogen and nitrogen.

b $N_2O_4(g) \rightleftharpoons 2NO_2(g)$

The standard free energy change for this reaction is

$$\begin{aligned}\Delta G^{\ominus}_{298} &= 2\Delta G^{\ominus}_{f,298}\,[NO_2(g)] - \Delta G^{\ominus}_{f,298}\,[N_2O_4(g)]\\ &= 2\times 51.3 - 97.8 = 4.8\,\text{kJ}\end{aligned}$$

From this

$$\lg K_p = \frac{4.8}{-5.69} = -0.844 \quad \text{and}$$

$$K_p = \frac{p^2_{NO_2eqm}}{p_{N_2O_4eqm}} = 0.14\,\text{atm at 298 K}$$

You may have investigated this equilibrium experimentally in section 12.2, in which case you should compare this with the results you obtained then.

2 *Heterogeneous equilibria involving a gas*

The equilibrium constant for the system

$$CaCO_3(s) \rightleftharpoons CaO(s) + CO_2(g)$$

is given by $K_p = p_{CO_2eqm}$, the partial pressures of $CaCO_3$ and CaO in the gas phase being constant and therefore incorporated in the equilibrium constant K_p.

The equilibrium pressure of CO_2 over $CaCO_3$ at 25 °C may be calculated as follows.

For the above reaction

$$\begin{aligned}\Delta G^{\ominus}_{298} &= \Delta G^{\ominus}_{f,298}[CO_2(g)] + \Delta G^{\ominus}_{f,298}[CaO(s)] - \Delta G^{\ominus}_{f,298}[CaCO_3(s)] \\ &= -395 + (-604) - (-1130) \\ &= 131 \text{ kJ}\end{aligned}$$

Now

$$\Delta G^{\ominus}_{298} = -5.69 \lg K_p = -5.69 \lg p_{CO_2 eqm}$$

so that

$$\lg p_{CO_2 eqm} = -23.0 \quad \text{and} \quad p_{CO_2 eqm} = 1.0 \times 10^{-23} \text{ atm}$$

As expected, calcium carbonate is highly stable at room temperature and pressure.

3 *Equilibrium constants for reactions between ions in solution*
Given the standard free energies of formation of the following ions

Ion	$\Delta G^{\ominus}_{f,298}$/kJ mol^{-1}
$Fe^{3+}(aq)$	-10.7
$Fe^{2+}(aq)$	-84.9
$Ag^{+}(aq)$	$+77.2$

find the equilibrium constant at 25 °C for the reaction

$$Ag^{+}(aq) + Fe^{2+}(aq) \rightleftharpoons Fe^{3+}(aq) + Ag(s)$$

(In dealing with ions, the standard free energy of formation of the hydrogen ion $\Delta G^{\ominus}_{f}[H^{+}(aq)] = 0$. This corresponds to the convention of regarding the standard potential of the hydrogen electrode as zero volts.)

The standard free energy $\Delta G^{\ominus}_{298}$ for the reaction is given by

$$\begin{aligned}\Delta G^{\ominus}_{298} &= \Delta G^{\ominus}_{f,298}[Fe^{3+}(aq)] + \Delta G^{\ominus}_{f,298}[Ag(s)] \\ &\quad - \Delta G^{\ominus}_{f,298}[Ag^{+}(aq)] - \Delta G^{\ominus}_{f,298}[Fe^{2+}(aq)] \\ &= -10.7 + 0 - 77.2 - (-84.9) \\ &= -3.0 \text{ kJ}\end{aligned}$$

Using this value in the equation

$$\Delta G^{\ominus}_{298} = -5.69 \lg K$$

we obtain $\lg K_c = 0.53$ and

$$K_c = \frac{[Fe^{3+}(aq)]}{[Ag^{+}(aq)]_{eqm}\,[Fe^{2+}(aq)]_{eqm}} = 3.4\ dm^3\ mol^{-1}$$

(Previous result from e.m.f. calculation – see experiments 17.3b and 17.3c – $K_c = 3.2\ dm^3\ mol^{-1}$)

Variation of the stabilities of compounds in the Periodic Table

In section 17.3 we saw that, since $\Delta G^{\ominus}$ is defined as

$$\Delta G^{\ominus} = -2.3\ RT \lg K,$$

the values of $\Delta G^{\ominus}$ correspond to values of K as in figure 17.3a and table 17.3a. Consequently the value of $\Delta G^{\ominus}_{f,298}$ for any compound is an indication of the *energetic* stability of that compound relative to formation from its elements in their standard states. A graph of $\Delta G^{\ominus}_{f,298}$ for a given set of compounds, such as the oxides of the elements, plotted against the atomic numbers of the elements, indicates the variation in energetic stability of the compounds across the Periodic Table.

The diagrams on the next few pages show the standard free energy of formation at 298 K of various compounds plotted against the atomic numbers of the elements forming these compounds.

Oxides

For the general reaction between an element M and oxygen, the following equations can be formulated:

1 $\frac{x}{y}M(s) + \frac{1}{2}O_2(g) \rightarrow \frac{1}{y}M_xO_y(s)$

2 $M(s) + \frac{y}{2x}O_2(g) \rightarrow \frac{1}{x}M_xO_y(s)$

Both equations allow a comparison to be made between the standard free energies of formation of various metal oxides, the first equation having been written in terms of one mole of oxygen atoms as the 'common unit', the second one referring to one mole of M atoms. The free energy spectrum, based on equation (2), for oxides* is given in figures 17.4a.

* Where metals have several oxidation numbers in forming oxides, the data given are those for the most common oxides.

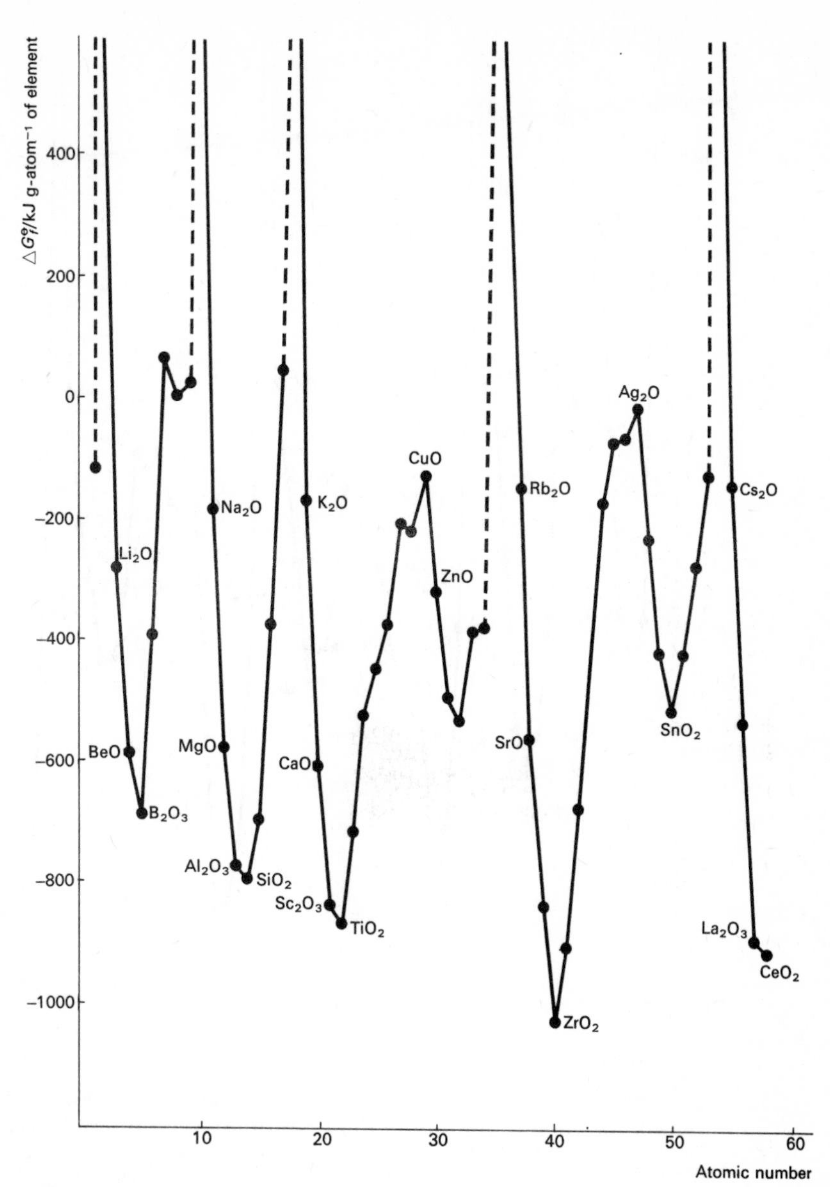

Figure 17.4a
Free energy spectrum of oxides at 298 K.

[General reaction: $M(s) + \frac{y}{2x}O_2(g) \rightarrow \frac{1}{x}M_xO_y(s)$]

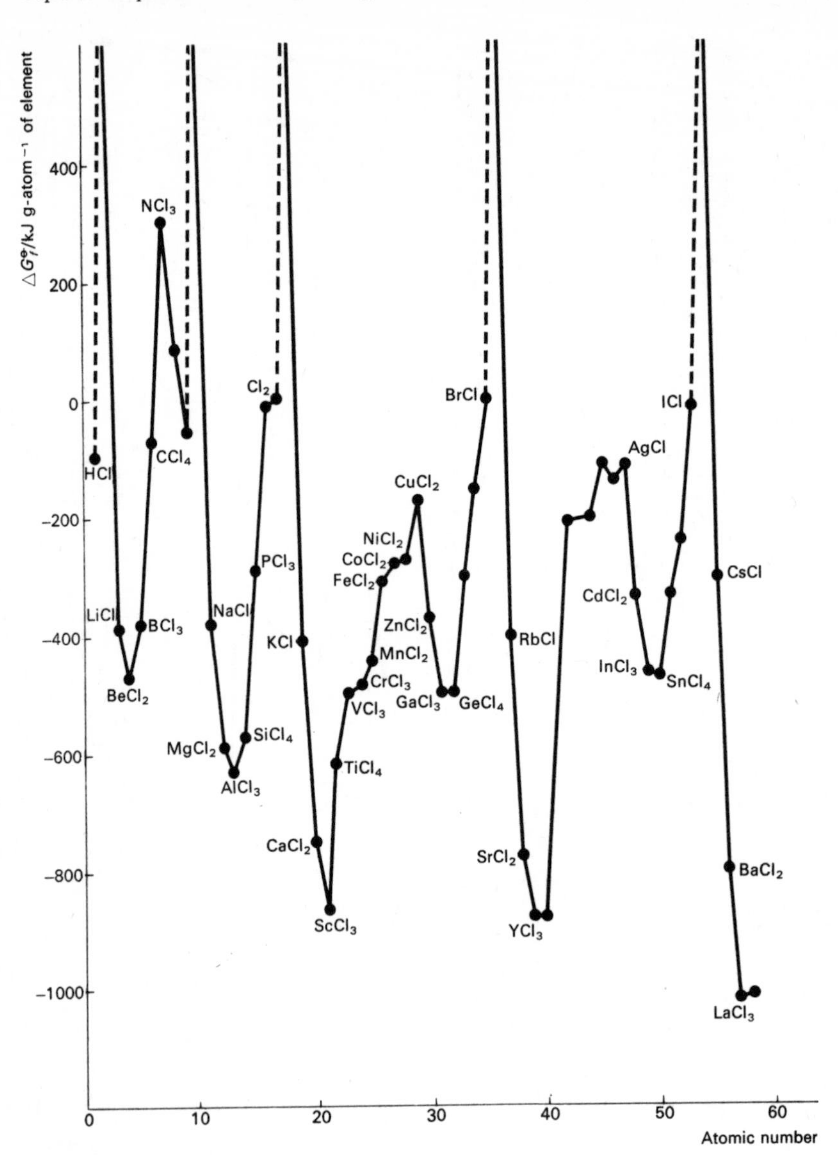

Figure 17.4b
Free energy spectrum of chlorides at 298 K.

[General reaction: $M(s)+\frac{x}{2}Cl_2(g) \rightarrow MCl_x(s)$]

Chlorides

A similar diagram for chlorides is given in figure 17.4b.
The general equation here is

$$M(s)+\frac{x}{2}\,Cl_2(g) \rightarrow M\,Cl_x(s).$$

Note that for these groups of compounds there is a systematic variation in the free energy of formation across the periodic classification: stabilities increase and decrease in accordance with the periods.

Transition metal oxides

You may like to look at some standard free energy data for the formation of transition metal oxides. Table 5 in the appendix to this topic contains the necessary information which is also plotted in figure 17.4c.

Hydrides of some elements

Compare the values of $\Delta G^{\ominus}_{f,298}$ for the hydrides of some elements, given in appendix table 6 with corresponding $\Delta H^{\ominus}_{f,298}$ values in appendix table 7. Comparison is most easily done graphically.

The relative stability of ions of different oxidation numbers

Figure 17.4d shows some standard free energies of formation of ions in aqueous solution for metals of the first transition series.

Information about the stability of one hydrated ion relative to another is obtained by calculating the standard free energy change for the (hypothetical) equation which describes the conversion of the one ion into the other. Consider for example the relative stabilities of copper(I) and copper(II) ions in aqueous solution.

1 $2Cu^{+}(aq) \rightarrow Cu^{2+}(aq)+Cu(s);\quad \Delta G^{\ominus}_{298} = -35.8\ \text{kJ}$

There are occasions when the above reaction is in principle reversed, e.g. when the $Cu^{+}(aq)$ ions are removed from the solution by precipitation with chloride ions, as in the preparation of copper(I) chloride.

2 $Cu^{+}(aq)+Cl^{-}(aq) \rightarrow Cu(Cl(s);\quad \Delta G^{\ominus}_{298} = -38.0\ \text{kJ}$

($\Delta G^{\ominus}_{f,298}[CuCl(s)] = -118.8$; $\Delta G^{\ominus}_{f,298}[Cl^{-}(aq)] = -131.2$
$\Delta G^{\ominus}_{f,298}[Cu^{+}(aq)] = +50.4\ \text{kJ mol}^{-1}$)

The free energy decrease per mole of Cu^{+} ions is larger in reaction (2) than in reaction (1). Hence, the formation of CuCl(s) is a feasible reaction.

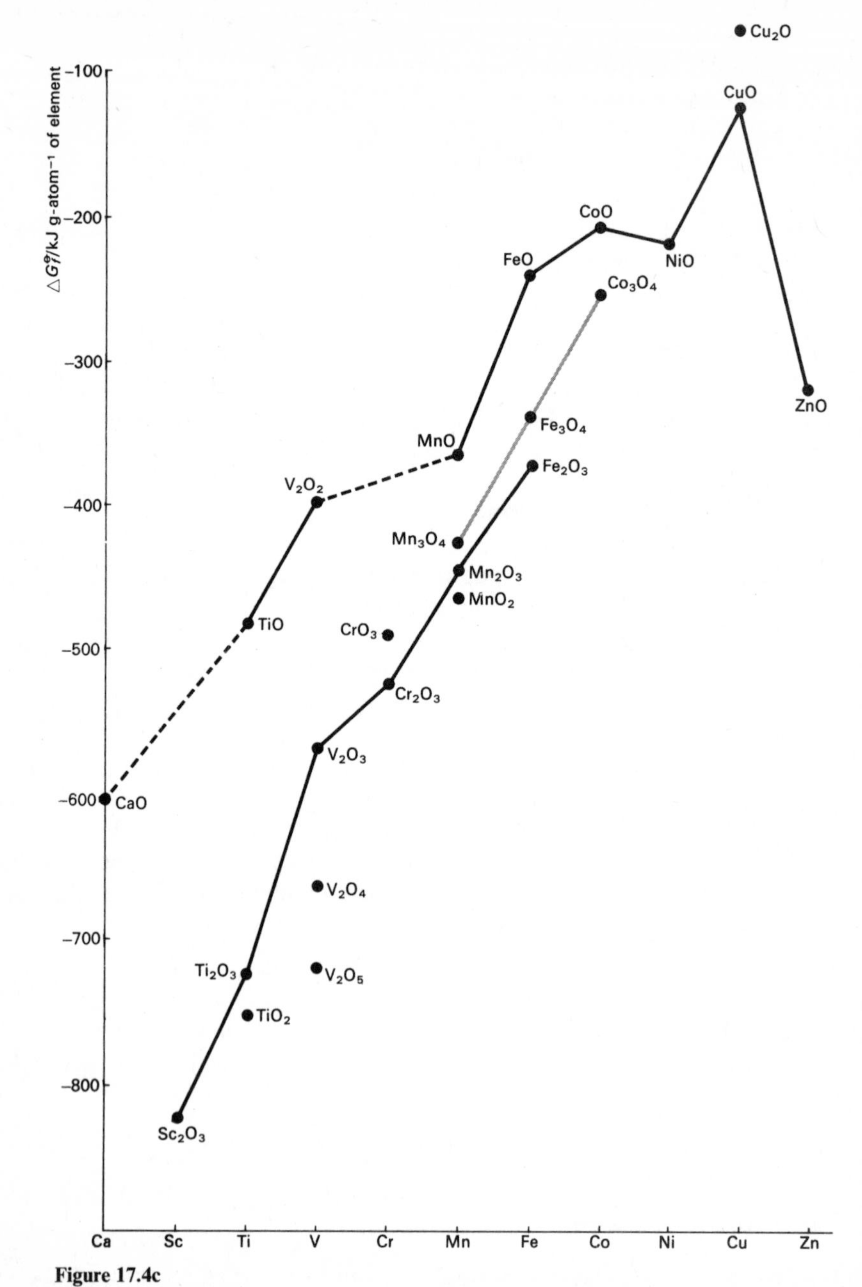

Figure 17.4c
Free energy diagram for metal oxides of first transition series.
[General reaction: $M(s)+\frac{y}{2x}O_2(g) \rightarrow \frac{1}{x}M_xO_y(s)$]

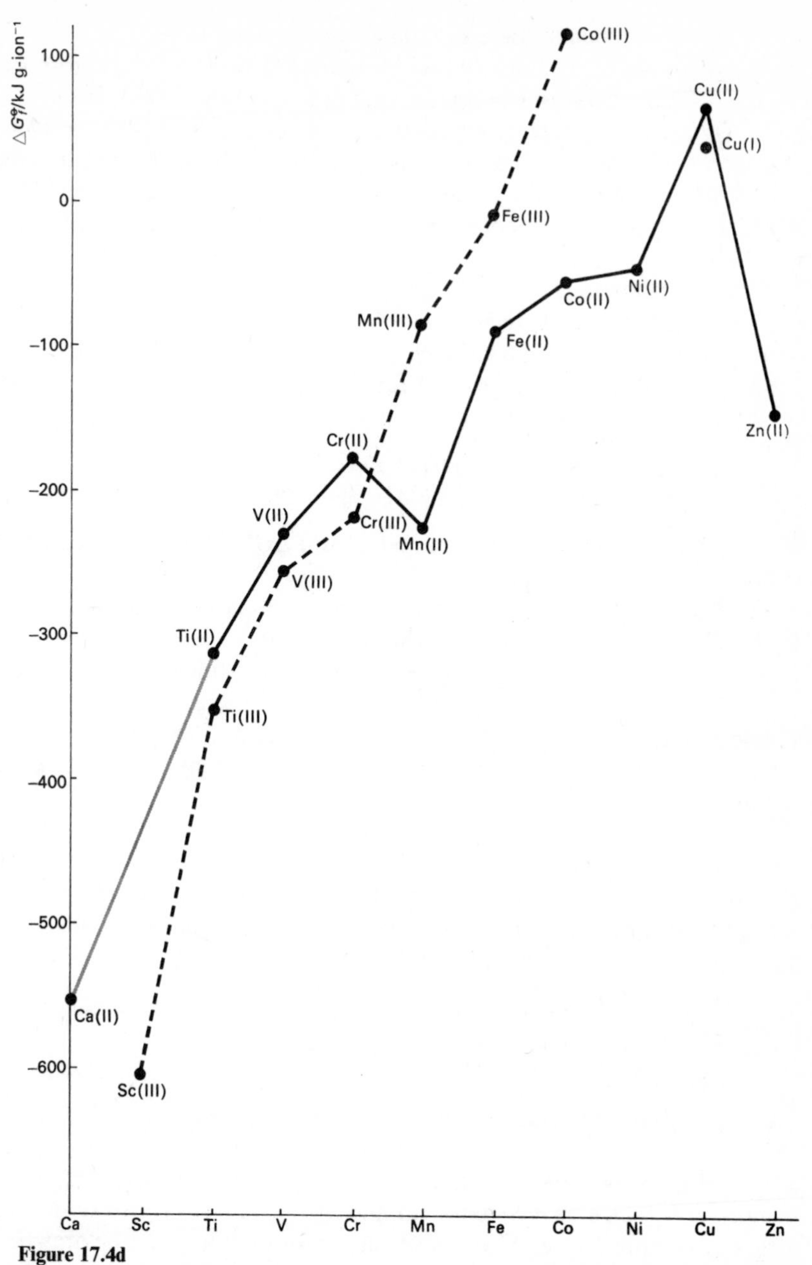

Figure 17.4d
Free energies of formation of ions in aqueous solutions for first transition metal series.

The extraction of metals – Ellingham diagrams

If we plot values of $\Delta G^\ominus$ for suitable reactions at various temperatures we can use the resulting diagrams to decide what processes for the extraction of metals are likely to work under given conditions. These diagrams (figure 17.4e is an example) show $\Delta G^\ominus$ values for reactions in which one mole of molecular oxygen at one atmosphere pressure combines with a pure element to form an oxide. These are examples of Ellingham diagrams, named after their inventor.

Example

Can zinc oxide be reduced by hydrogen (a) at 500 °C (b) at 1500 °C?

1 Write an equation for the proposed reduction.

Problem: what is the value of $\Delta G^\ominus$ for this reaction?

2 Consider the reactions at 500 °C

$$2Zn(s)+O_2(g) \rightarrow 2ZnO(s);\ \Delta G_1^\ominus$$
$$2H_2(g)+O_2(g) \rightarrow 2H_2O(g);\ \Delta G_2^\ominus$$

Write down appropriate values for $\Delta G_1^\ominus$ and $\Delta G_2^\ominus$ by reading them from the graph.

3 What must you do with the equations and $\Delta G^\ominus$ values in (2) to obtain the equation you wrote in (1) and the value of $\Delta G^\ominus$ for it?

4 Would you expect hydrogen to reduce zinc oxide at 500 °C?

5 Read off values of $\Delta G_1^\ominus$ and $\Delta G_2^\ominus$ at 1500 °C and repeat the calculation. Would you expect hydrogen to reduce zinc oxide at 1500 °C?

6 Above what temperature would you expect hydrogen to become a suitable reducing agent for zinc oxide?

Would carbon do instead of hydrogen?

Hydrogen is expensive and so is working at high temperatures. Can we use carbon (a cheaper material) at a lower temperature? Three possible oxidation reactions to be considered are

$$2C(s)+O_2(g) \rightarrow 2CO(g)$$
$$C(s)+O_2(g) \rightarrow CO_2(g)$$
$$2CO(g)+O_2(g) \rightarrow 2CO_2(g)$$

The appropriate Ellingham diagram is given as figure 17.4f.

Taking the first two reactions only, below 750°C it would seem that $C \rightarrow CO_2$ is more active as a reducing system and above this temperature the reaction $C \rightarrow CO$ is more active. A composite curve (dotted line) can be drawn.

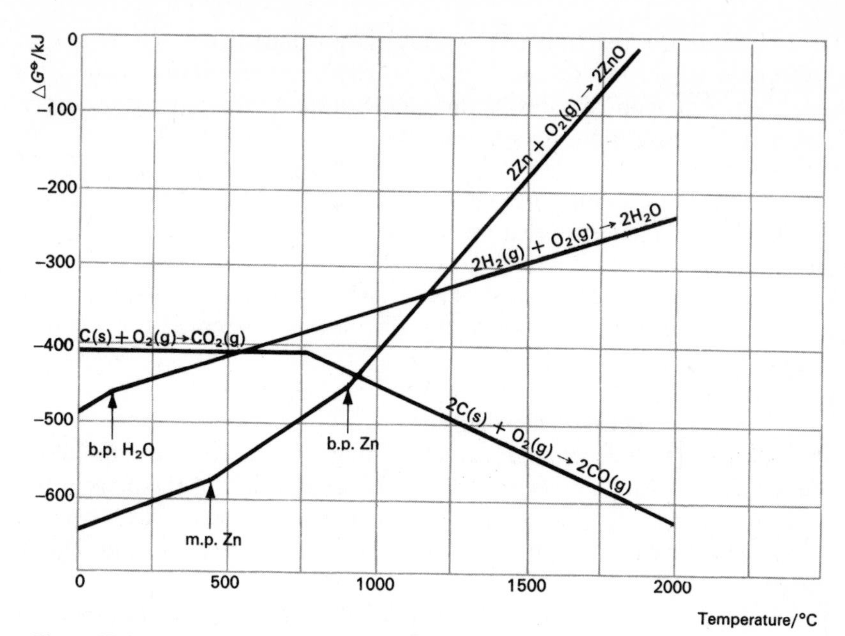

Figure 17.4e

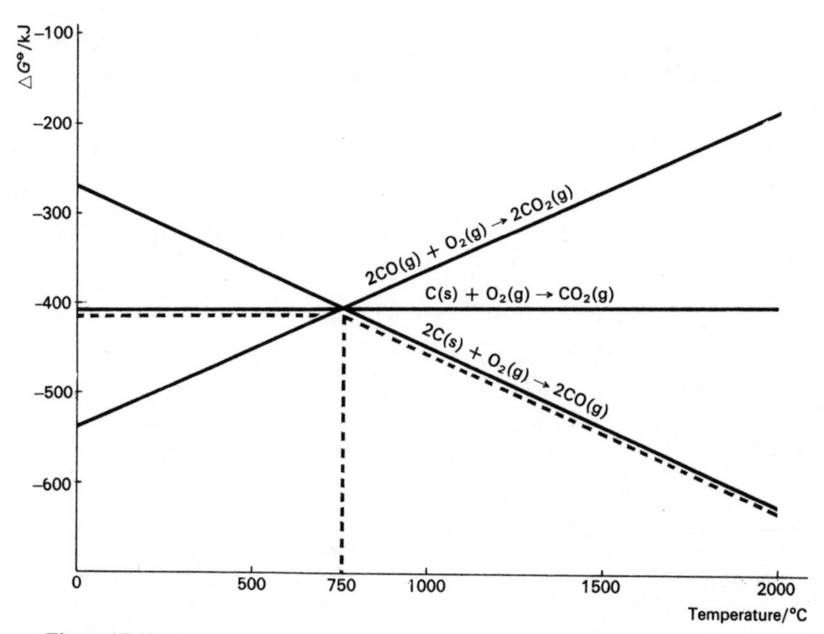

Figure 17.4f

The line for the reaction $CO \rightarrow CO_2$ shows that below 750 °C if CO were present it would be the most effective reducing agent. But carbon will not supply carbon monoxide below this temperature, so the composite curve shows the most effective reducing system.

Now examine the composite curve in relation to the zinc/zinc oxide curve.

1 Above what temperature would you expect carbon to be an effective reducing agent for zinc oxide?

2 In what phase (physical state) is zinc at this temperature?

3 Do you think the state of the zinc at this temperature will facilitate or hinder the reduction process?

Kinetic and energetic stabilities

In Topic 7 (Energy changes and bonding) we made a careful distinction between *kinetic* and *energetic* stability. In section 17.1 of the present topic we referred again to the importance of kinetic energy barriers in preventing the attainment of equilibrium. The next two examples illustrate this point once more.

1 Cane sugar has the formula $C_{12}H_{22}O_{11}$ and is one of the most common chemicals found in the household. When stored, it is almost always exposed to atmospheric oxygen, and thus appears to be stable with respect to the latter. Assess the stability of cane sugar in atmospheric oxygen on energetic grounds.

Data required

	$\Delta G^{\ominus}_{f,298}/\text{kJ mol}^{-1}$
Cane sugar $C_{12}H_{22}O_{11}(s)$	-1544
$CO_2(g)$	-394
$H_2O(l)$	-230

Assume that the reaction between cane sugar and oxygen would proceed as follows:

$$C_{12}H_{22}O_{11}(s) + 12O_2(g) \rightarrow 12CO_2(g) + 11H_2O\,(l)$$

Calculate $\Delta G^{\ominus}_{298}$ for this reaction.

In view of this large negative free energy value, cane sugar might be expected to react spontaneously with oxygen to form carbon dioxide and water. Thus cane sugar is *energetically* unstable with respect to oxygen.

The activation energy for this reaction is, however, very considerable. (See also Topic 14, Introduction.)

2 Figure 17.4g shows the standard free energies of formation, at 298 K, of the initial eight homologues of the alkane series and of ethylene.

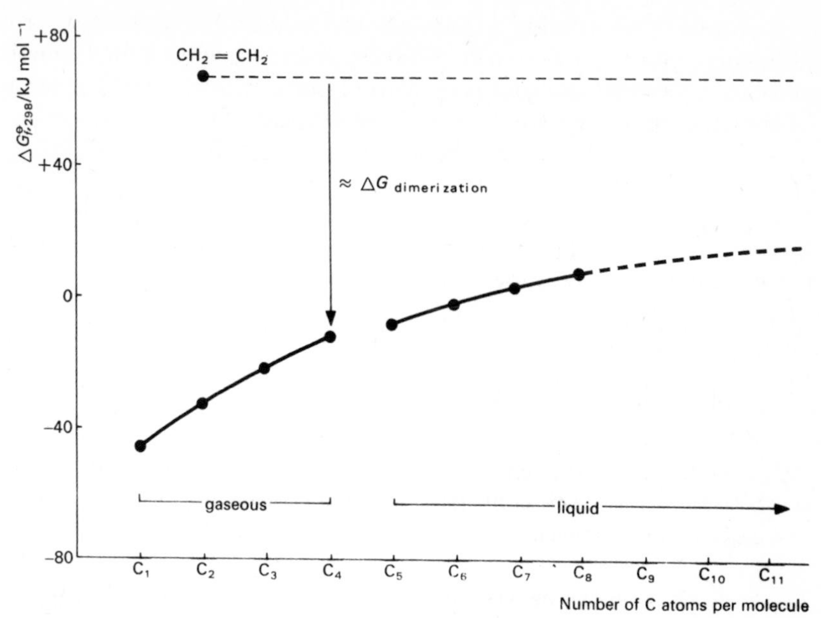

Figure 17.4g

The general shape of the $\Delta G^{\ominus}_{f,298}$ curve for the alkanes suggests that as the number of carbon atoms increases, $\Delta G^{\ominus}_{f,298}$ approaches asymptotically a limiting value which lies below the $\Delta G^{\ominus}_{f,298}$ value of C_2H_4. In consequence, the polymerization of ethylene, represented by the overall equation

$$n\,CH_2{=}CH_2(g) \rightarrow (-CH_2-CH_2-)_n$$

should take place at room temperature.

It is, however, possible to prepare and store ethylene at room temperature for an indefinitely long period of time without polymerization taking place. Only through the use of catalysts can the polymerization be brought about at a suitable rate.

Energetics in life processes

This is a field in which our knowledge about energetic relationships is very limited indeed. Yet life itself depends on the continuous supply to organisms of energy in the form of foodstuffs. And there can be little doubt that the human body is among other things a highly efficient machine driven by chemical fuels.

Biochemists have studied a considerable number of metabolic processes occurring in plants and animals. All these seem to involve highly complicated chemical processes, and so it is perhaps reasonable to assume that the supply of energy to organisms will ultimately be explicable in terms of bond making and bond breaking, just as in simpler chemical systems.

But do these processes obey the general laws of energetics? Are they controlled by the same factors which have been seen to apply to the reactions chemists can carry out in their laboratories?

The present state of our knowledge makes it impossible to give final answers to questions of this sort and any provisional answers we offer necessarily involve a measure of speculation.

Probably the most studied and best known biochemical reaction is photosynthesis, which we now consider briefly from the point of view of the energy changes that are involved.

The basic reaction can be written

$$6CO_2(g) + 6H_2O(l) \rightarrow C_6H_{12}O_6(s) + 6O_2(g)$$

The free energy change for this process is approximately +2900 kJ, so that the reaction is energetically not feasible. Yet, it is a well established process of nature. Biochemists have discovered other natural processes which are accompanied by an increase in free energy.

One approach to an explanation of this is to consider more carefully what constitutes the *system* in this instance. As is well known, the above reaction will not happen in the absence of light, that is, in a system isolated with respect to light energy. The energy balance may however add up rather differently if the energy associated with the light entering the photo-activated system is taken into account.

For photosynthesis such a calculation can be done as follows:

Apart from sugars of formula $C_6H_{12}O_6$, other carbohydrates are produced by the photosynthesis reaction. For the generalized reaction

$$CO_2(g) + H_2O(l) \rightarrow CH_2O(s) + O_2(g)$$

approximately +480 kJ are required per mole of carbon atoms. The empirical formula (that is, the ratio of the number of atoms present in the molecule) of the sugar is used in this equation because the free energy change is nearly independent of the degree of polymerization of the carbohydrate formed. This quantity of energy is supplied by photons. It is known that the absorption of light energy by plants is facilitated by the chlorophyll which absorbs the yellow-red light of wavelength 600 nm ($= 6 \times 10^{-7}$ m).

The energy of light of this wavelength can be calculated as follows: the energy of a photon is given by the Planck equation

$$E = h\upsilon$$

where E = energy in 10^{-7} joules, h the Planck constant (6.6×10^{-34} J s) and υ the frequency of the light in s^{-1}. The wavelength of the light, λ, is equal to c/υ, c being the velocity of light (3×10^8 m s^{-1}); thus $\upsilon = c/\lambda$

For one mole of photons:

$$E = \frac{6.02 \times 10^{23} \times 6.6 \times 10^{-34} \times 3 \times 10^8}{6 \times 10^{-7}} = 200 \times 10^3 \text{ joules} = 200 \text{ kJ}$$

The conclusion may thus be drawn that approximately 3 moles at least of photons of wavelength 600 nm are required per mole of carbon atoms in the photosynthesis process.

Problems

* indicates that the *Book of Data* is required.

1 Given that for the reaction

$$N_2O_4(g) \rightleftharpoons 2NO_2(g)$$

at 600 K, $\Delta H^{\ominus} = 57.1$ kJ and $\lg K_p = 4.25$,

i calculate the value of K_p at 600 K

ii hence calculate a value for K_p at 1000 K.

iii What assumptions do you make in answering (*ii*)?

2 The following are the data for the 'water gas reaction'.

$$H_2O(g) + C(s) \rightleftharpoons H_2(g) + CO(g)$$

T/K	$\Delta H^{\ominus}_T$/kJ	$\Delta G^{\ominus}_T$/kJ
1300	+146.9	−57.6

i Calculate (a) $\lg K_p$

(b) K_p at 1300 K

ii Calculate K_p at 1500 K

iii At which of these temperatures is the equilibrium yield of $H_2(g)$ and $CO(g)$ greater?

3 The equilibrium constant for the reaction

$$Ag^+(aq) + Fe^{2+}(aq) \rightleftharpoons Fe^{3+}(aq) + Ag(s)$$

as calculated from $E^\ominus$ values, is 3.2 dm^3 mol^{-1} at 25 °C.

i Use the standard heats of formation of the aqueous ions involved to calculate the standard enthalpy change for this reaction at 25 °C.

ii Use the equilibrium constant to calculate the standard free energy change for this reaction at 25 °C.

4 The graph below (figure 17.5) represents the variation with temperature of the equilibrium constant for the vapour phase dissociation of dimeric acetic acid,

$$(CH_3CO_2H)_2(g) \rightleftharpoons 2CH_3CO_2H(g)$$

Calculate the values of (*i*) $\Delta H^\ominus$ and (*ii*) $\Delta G^\ominus$ for the above reaction at 150 °C.

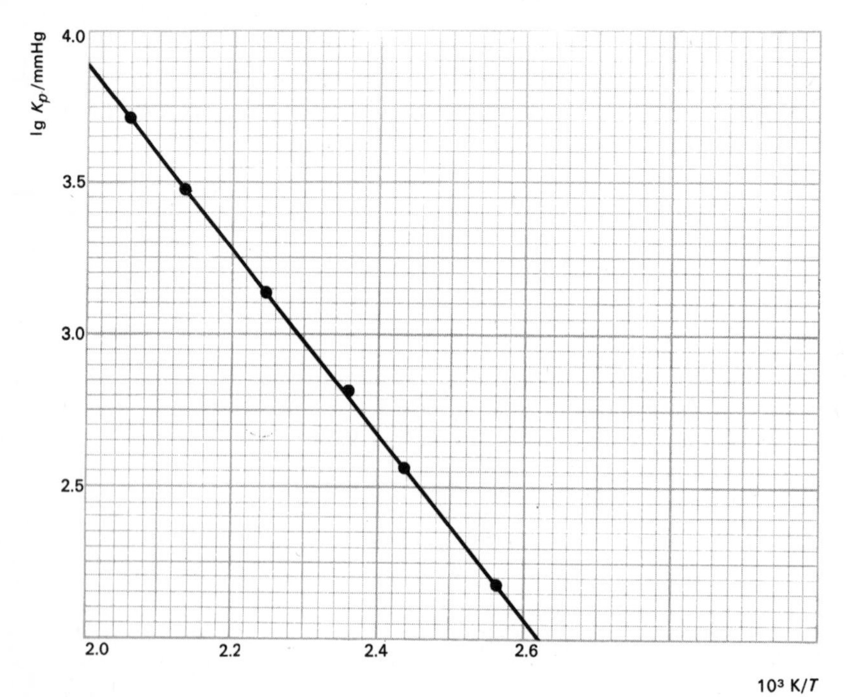

Figure 17.5
Data from Johnson, E. W. and Nash, L. K. (1950) J. Amer. Chem. Soc. **72**, *547.*

***5** When an aqueous solution of bromine is added to a solution of potassium iodide, the following reaction takes place

$$\tfrac{1}{2}Br_2(aq)+I^-(aq) \rightarrow \tfrac{1}{2}I_2(aq)+Br^-(aq)$$

i Write down a cell diagram for an electrochemical cell in which this reaction can also be carried out.

ii From the *Book of Data* find the standard e.m.f. of your cell.

iii What is K_c for the above reaction?

***6** The oxidation of $Fe^{2+}(aq)$ ions by $MnO_4^-(aq)$ ions proceeds according to the equation

$$5Fe^{2+}(aq)+MnO_4^-(aq)+8H^+(aq) \rightarrow 5Fe^{3+}(aq)+Mn^{2+}(aq)+4H_2O(l)$$

for which

$$K_c = \frac{[Fe^{3+}(aq)]^5_{eqm}[Mn^{2+}(aq)]_{eqm}}{[MnO_4^-(aq)]_{eqm}[Fe^{2+}(aq)]^5_{eqm}[H^+(aq)]^8_{eqm}}$$

i Write two half-equations, each one for a distinct half-cell reaction.

ii Write a cell diagram for a suitable electrochemical cell in which this reaction could be carried out.

iii Calculate the standard e.m.f. of this cell (at 298 K).

iv What is the value of K_c at 298 K?

***7**

i Use tables to write down the standard free energy of formation of methane gas, CH_4.

ii Does this value suggest that the reaction

$$C(s)+2H_2(g) \rightarrow CH_4(g)$$

would, from considerations of free energy, be expected to take place?

iii In your experience does such a reaction take place at room temperature and atmospheric pressure?

iv How do you explain any difference in your answers to (*ii*) and (*iii*) above?

***8** Use tables of free energies of formation to compare the stabilities of

i iron(II) oxide and iron(III) oxide relative to their formation from the elements.

ii the stability of iron(II) oxide relative to atmospheric oxidation to iron(III) oxide.

9 The relevant equation and data for the formation of the diamminosilver ion are

$$Ag^+(aq) + 2NH_3(aq) \rightarrow [Ag(NH_3)_2]^+(aq)$$
$$\Delta G^{\ominus}_{f,298}[Ag^+(aq)] = +1.8 \text{ kJ mol}^{-1}$$
$$\Delta G^{\ominus}_{f,298}[NH_3(aq)] = +26.6 \text{ kJ mol}^{-1}$$
$$\Delta G^{\ominus}_{f,298}[[Ag(NH_3)_2]^+(aq)] = -17.0 \text{ kJ mol}^{-1}$$

i Calculate $\Delta G^{\ominus}_{298}$ for this reaction.

ii Is the stability of the diamminosilver ion greater or less than that of the ordinary hydrated silver ion?

iii Calculate K_c for the above reaction.

iv Invent an experimental method for determining K_c for this reaction

a by means of an electrochemical cell

b by some other means.

10 The relevant equation and data for the dissolution of solid aluminium hydroxide in alkaline solution are

$$Al(OH)_3(s) + OH^-(aq) \rightarrow Al(OH)_4^-(aq)$$
$$\Delta G^{\ominus}_{f,298}[Al(OH)_4^-(aq)] = -1310 \text{ kJ mol}^{-1}$$
$$\Delta G^{\ominus}_{f,298}[OH^-(aq)] = -159 \text{ kJ mol}^{-1}$$
$$\Delta G^{\ominus}_{f,298}[Al(OH)_3(s)] = -1142 \text{ kJ mol}^{-1}$$

i Calculate $\Delta G^{\ominus}_{298}$ for the above reaction.

ii Hence calculate K_c.

iii On the basis of this calculation would you expect aluminium hydroxide to dissolve in molar aqueous alkali?

Appendix

Data for Topic 17

Table 1

Equilibrium constants (K_p) of some gaseous reactions at various temperatures
(*Values of K_p calculated from partial pressures in atmospheres*)

Temperature T/K	$\frac{1}{T}$/K^{-1}	K_p	lg K_p
Reaction: $N_2O_4(g) \rightleftharpoons 2NO_2(g)$; $\Delta H^{\ominus}_{298} = +58.2$ kJ			
298	3.36×10^{-3}	1.15×10^{-1}	−0.94
350	2.86×10^{-3}	3.89	+0.59
400	2.50×10^{-3}	4.79×10^{1}	+1.68
450	2.22×10^{-3}	3.47×10^{2}	+2.54
500	2.00×10^{-3}	1.70×10^{3}	+3.23
550	1.82×10^{-3}	6.03×10^{3}	+3.78
600	1.67×10^{-3}	1.78×10^{4}	+4.25
Reaction: $H_2O(g) + C(s) \rightleftharpoons H_2(g) + CO(g)$; $\Delta H^{\ominus}_{298} = +131$ kJ			
298	3.36×10^{-3}	1.00×10^{-16}	−16.0
500	2.00×10^{-3}	2.52×10^{-7}	−6.60
700	1.43×10^{-3}	2.82×10^{-3}	−2.55
800	1.25×10^{-3}	5.37×10^{-2}	−1.27
900	1.11×10^{-3}	5.75×10^{-1}	−0.24
1000	1.00×10^{-3}	3.72	+0.57
1100	9.09×10^{-4}	1.70×10^{1}	+1.23
1200	8.34×10^{-4}	6.60×10^{1}	+1.82
1300	7.69×10^{-4}	2.04×10^{2}	+2.31
Reaction: $H_2(g) + CO_2(g) \rightleftharpoons H_2O(g) + CO(g)$; $\Delta H^{\ominus}_{298} = +41$ kJ			
298	3.36×10^{-3}	1.00×10^{-5}	−5.00
500	2.00×10^{-3}	7.76×10^{-3}	−2.11
700	1.43×10^{-3}	1.23×10^{-1}	−0.91
800	1.25×10^{-3}	2.88×10^{-1}	−0.54
900	1.11×10^{-3}	6.03×10^{-1}	−0.22
1000	1.00×10^{-3}	9.55×10^{-1}	−0.02
1100	9.09×10^{-4}	1.45	+0.16
1200	8.34×10^{-4}	2.10	+0.32
1300	7.69×10^{-4}	2.82	+0.45

Temperature T/K	$\frac{1}{T}$/K^{-1}	K_p	lg K_p
Reaction: $N_2(g)+3H_2(g) \rightleftharpoons 2NH_3(g)$; $\Delta H^{\ominus}_{298} = -92$ kJ			
298	3.36×10^{-3}	6.76×10^{5}	+5.83
400	2.50×10^{-3}	4.07×10^{1}	+1.61
500	2.00×10^{-3}	3.55×10^{-2}	−1.45
600	1.67×10^{-3}	1.66×10^{-3}	−2.78
700	1.43×10^{-3}	7.76×10^{-5}	−4.11
800	1.25×10^{-3}	6.92×10^{-6}	−5.16
900	1.11×10^{-3}	1.00×10^{-6}	−6.00
Reaction: $2SO_2(g)+O_2(g) \rightleftharpoons 2SO_3(g)$; $\Delta H^{\ominus}_{298} = -197$ kJ			
298	3.36×10^{-3}	4.0×10^{24}	24.60
500	2.00×10^{-3}	2.5×10^{10}	10.40
700	1.43×10^{-3}	3.0×10^{4}	4.48
1100	9.09×10^{-4}	1.3×10^{-1}	−0.89
Reaction: $Ag_2CO_3(s) \rightleftharpoons Ag_2O(s)+CO_2(g)$; $\Delta H^{\ominus}_{298} = +81.7$ kJ			
298	3.36×10^{-3}	3.16×10^{-6}	−5.5
350	2.86×10^{-3}	3.98×10^{-4}	−3.4
400	2.50×10^{-3}	1.41×10^{-2}	−1.85
450	2.22×10^{-3}	1.86×10^{-1}	−0.73
500	2.00×10^{-3}	1.48	+0.17
550	1.82×10^{-3}	8.91	+0.95
600	1.67×10^{-3}	6.31×10^{1}	+1.8

Table 2
Energy data for some non-ionic reactions at various temperatures
(*All energy data in* kJ)

Temperature T/K	$\Delta H^{\ominus}_{T}$	$\Delta G^{\ominus}_{T}$	lg K_p
Reaction: $Ag_2CO_3(s) \rightleftharpoons Ag_2O(s)+CO_2(g)$			
298	81.7	31.4	−5.5
350	81.2	23.0	−3.4
400	80.3	14.2	−1.85
450	79.9	6.3	−0.73
500	79.5	−1.7	+0.17
550	78.7	−10.0	+0.95
600	78.3	−20.5	+1.8

Temperature T/K	$\Delta H_T^{\ominus}$	$\Delta G_T^{\ominus}$	$\lg K_p$
Reaction: $CaCO_3(s) \rightleftharpoons CaO(s) + CO_2(g)$			
298	177.9	130.0	−22.8
500	177.5	97.5	−10.2
700	177.0	65.3	−4.9
800	177.0	49.8	−3.3
900	177.0	33.9	−2.0
1000	176.5	18.0	−0.9
1100	176.5	2.1	−0.1
1200	176.0	−13.8	+0.6
1300	176.0	−29.7	+1.2
Reaction: $N_2O_4(g) \rightleftharpoons 2NO_2(g)$			
298	58.23	4.80	−0.94
350	57.86	−3.93	+0.59
400	57.69	−12.93	+1.68
450	57.57	−21.88	+2.54
500	57.40	−30.88	+3.23
550	57.14	−39.87	+3.78
600	57.07	−48.82	+4.25
Reaction: $N_2(g) + 3H_2(g) \rightleftharpoons 2NH_3(g)$			
298	−92.1	−33.3	+5.83
400	−96.9	−12.3	+1.61
500	−101.3	13.9	−1.45
600	−105.8	31.9	−2.78
700	−110.2	55.1	−4.11
800	−114.7	79.1	−5.16
900	−119.1	103.3	−6.00
Reaction: $H_2O(g) + C(s) \rightleftharpoons H_2(g) + CO(g)$			
298	131.3	91.3	−16.0
500	134.4	63.3	−6.60
700	137.6	34.2	−2.55
800	139.1	19.5	−1.27
900	140.7	4.2	−0.24
1000	142.3	−11.0	+0.57
1100	143.9	−26.0	+1.23
1200	145.4	−41.8	+1.82
1300	146.9	−57.6	+2.32

Temperature T/K	$\Delta H^{\ominus}_{T}$	$\Delta G^{\ominus}_{T}$	$\lg K_p$
Reaction: $H_2(g) + CO_2(g) \rightleftharpoons H_2O(g) + CO(g)$			
298	41.2	28.5	−5.00
500	40.5	20.2	−2.11
700	39.9	12.2	−0.91
800	39.5	8.2	−0.54
900	39.1	4.2	−0.22
1000	38.8	0.3	−0.02
1100	38.4	−3.5	+0.16
1200	38.1	−7.4	+0.32
1300	37.8	−11.1	+0.45

Table 3
Some energy changes and their relation to equilibrium constants and volume changes for selected reactions at 298 K
(*Energy data in* kJ*; equilibrium constants in appropriate units; volume changes in cubic decimetres*)

Reaction	**ΔV**	$\Delta H^{\ominus}_{298}$	$\Delta G^{\ominus}_{298}$	K_{298}	**lg** K_{298}
Type of reaction: GAS + GAS → GAS					
$CH_4(g) + 2O_2(g) \rightarrow CO_2(g) + 2H_2O(g)$	0	−802	−801	10^{140}	140
$H_2(g) + Cl_2(g) \rightarrow 2HCl(g)$	0	−184	−190	10^{33}	33
$CCl_4(g) + 2H_2O(g) \rightarrow CO_2(g) + 4HCl(g)$	+44.8	−173	−255	10^{45}	45
$N_2O_4(g) \rightarrow 2NO_2(g)$	+22.4	+58.2	+4.8	1.43×10^{-1}	−0.844
$N_2(g) + 3H_2(g) \rightarrow 2NH_3(g)$	−44.8	−92.0	−33.4	9.5×10^{5}	5.87
$C_2H_4(g) + H_2(g) \rightarrow C_2H_6(g)$	−22.4	−137	−119	7.9×10^{20}	20.9
Type of reaction: SOLID + GAS → SOLID					
$CaO(s) + CO_2(g) \rightarrow CaCO_3$	−22.4	−178	−130	10^{23}	23
$2Cu(s) + O_2(g) \rightarrow 2CuO(s)$	−22.4	−310	−254	10^{45}	45
Type of reaction: SOLID + GAS → GAS					
$C(s) + O_2(g) \rightarrow CO_2(g)$	0	−393	−394	10^{69}	69
$C(s) + H_2O(g) \rightarrow CO(g) + H_2(g)$	+22.4	+131	+91.3	10^{-16}	−16
Type of reaction: SOLID + GAS → SOLID + GAS					
$MgO(s) + CO(g) \rightarrow Mg(s) + CO_2(g)$	0	+319	+313	10^{-55}	−55
$ZnO(s) + CO(g) \rightarrow Zn(s) + CO_2(g)$	0	+65.0	+61.1	2×10^{-11}	−10.7
Type of reaction: SOLID + SOLID → SOLID					
$Cu(s) + S(s) \rightarrow CuS(s)$	0	−48.5	−49.0	3.98×10^{8}	8.6
$2Al(s) + Cr_2O_3(s) \rightarrow Al_2O_3(s) + 2Cr(s)$	0	−541	−530	10^{93}	93
$FeO(s) + Fe_2O_3(s) \rightarrow Fe_3O_4(s)$	0	−28.4	−28.9	1.14×10^{5}	5.1

Table 4
Some energy data, $E^{\ominus}_{298}$ values, and equilibrium constants for selected reactions in solution
(*Energy data in* kJ; *equilibrium constants in appropriate units; standard redox potentials in volts*)

Reaction	$\Delta H^{\ominus}_{298}$	$\Delta G^{\ominus}_{298}$	$E^{\ominus}_{298}$	$K_{c(298)}$
$Cu^{2+}(aq) + Zn(s) \rightarrow Cu(s) + Zn^{2+}(aq)$	−217	−212	+1.10	10^{37}
$Zn(s) + 2H^{+}(aq) \rightarrow Zn^{2+}(aq) + H_2(g)$	−152	−147	+0.76	10^{26}
$Pb^{2+}(aq) + Zn(s) \rightarrow Zn^{2+}(aq) + Pb(s)$	−154	−123	+0.64	10^{21}
$2Ag^{+}(aq) + Cu(s) \rightarrow 2Ag(s) + Cu^{2+}(aq)$	−147	−89	+0.46	10^{16}
$2Tl^{+}(aq) + Zn(s) \rightarrow 2Tl(s) + Zn^{2+}(aq)$	−164	−82	+0.42	10^{15}
$Cu^{2+}(aq) + Pb(s) \rightarrow Cu(s) + Pb^{2+}(aq)$	−62.8	−89.3	+0.47	10^{16}
$Tl(s) + H^{+}(aq) \rightarrow Tl^{+}(aq) + \frac{1}{2}H_2(g)$	+5.9	−31.8	+0.34	4×10^{5}

Table 5
Standard free energies of formation for some oxides of the first row transition series metals
($\Delta G_{f,298}$/kJ *mol*$^{-1}$)

	General formula								
	M_2O	MO	M_2O_2	M_2O_3	MO_2	M_2O_4	M_3O_4	M_2O_5	MO_3
(Ca)	—	−604.2	—	—	—	—	—	—	—
Sc	—	—	—	−1663	—	—	—	—	—
Ti	—	−482.4	—	−1426	−858.9	—	—	—	—
V	—	—	−905.8	−1252	—	−1446	—	−1699	—
Cr	—	—	—	−1046.8	—	—	—	—	−495.8
Mn	—	−363.2	—	−893.3	−466.1	—	−1280.3	—	—
Fe	—	−244.3	—	−741.0	—	—	−1014.2	—	—
Co	—	−213.4	—	—	—	—	−758.9	—	—
Ni	—	−216.3	—	—	—	—	—	—	—
Cu	−146.4	−127.2	—	—	—	—	—	—	—
Zn	—	−318.2	—	—	—	—	—	—	—

Table 6
Standard free energies of formation for some hydrides ($\Delta G^{\ominus}_{f,298}$/kJ *mol*$^{-1}$)

LiH	BeH_2	B_2H_6	CH_4	NH_3	H_2O	HF
−70.0	—	+82.8	−50.8	−16.7	−229	−273
NaH	MgH_2	AlH_3	SiH_4	PH_3	H_2S	HCl
−38.9	—	—	+56.9	+13.4	−33.6	−95.3
KH	CaH_2	Ga_2H_6	GeH_4	AsH_3	H_2Se	HBr
−22.2	−150	—	—	+68.9	+64.0	−53.2
RbH	SrH_2	InH_3	SnH_4	SbH_3	H_2Te	HI
−30.5	−137	—	—	+148	+130	+2.1
CsH	BaH_2	TlH_3	PbH_4	BiH_3		
−30.5	−132	—	—	—		
←SOLIDS→		←		GASES		→

Table 7
Standard enthalpies of formation for some hydrides
($\Delta H^{\ominus}_{f,298}$/kJ *mol*$^{-1}$ *to nearest integer*)

LiH	BeH_2	B_2H_6	CH_4	NH_3	H_2O	HF
−90	—	+31	−75	−46	−242	−271
NaH	MgH_2	AlH_3	SiH_4	PH_3	H_2S	HCl
−57	—	—	+34	+5	−21	−92
KH	CaH_2	Ga_2H_6	GeH_4	AsH_3	H_2Se	HBr
−60	−189	—	—	+66	+84	−36
RbH	SrH_2	InH_3	SnH_4	SbH_3	H_2Te	HI
−50	−176	—	—	+145	+75	+26
CsH	BaH_2	TlH_3	PbH_4	BiH_3		
−80	−171	—	—	—		
←SOLIDS→		←		GASES		→

Topic 18

Carbon compounds with large molecules

This topic considers some of the large molecules whose size depends on the ability of carbon atoms to link up in very long chains. If there are more than about a thousand carbon atoms in a single molecule it can be considered a giant or 'macromolecule' ('macro': large or long).

The first section of the topic is about detergents. Soapy detergents are themselves much too small to qualify as macromolecules but they are derived from *lipids*, which are much larger molecules. An experimental study will show how carbon chain length can influence properties, and a simple comparison of different types of detergents will reveal how synthetic compounds can be used as alternatives to naturally occurring materials.

The theme of synthetic materials is continued later in the topic with a study of some plastics. Plastics are of increasing importance as alternatives to naturally occurring materials such as wood and wool, but they also have properties which make them unique and not merely substitute materials.

The topic also considers an important class of naturally occurring macromolecules, the *proteins*. Proteins are built up from sub-units of amino acids.

There are two other important classes of organic macromolecules, *polysaccharides* and *nucleic acids*.

Polysaccharides are built up from glucose (see figure 18.A) and similar small molecules. They are used by plants and animals mainly as structural materials and as food reserves. Thus *cellulose*, a polymer of as many as five thousand glucose units, forms the framework for cells in plant tissue; the shell of a crab and the outer covering of a beetle are made of *chitin*, another polysaccharide. The main food reserve of plants is *starch* which forms up to 80 per cent of the weight of seeds. Starch molecules consist of up to three thousand sub-units of glucose.

The other class of macromolecules which is described only briefly here is the nucleic acids. The main chain of a nucleic acid consists of sugar molecules linked together by triphosphate groups, which are also responsible for the acidity of the molecule. Each sugar molecule is combined with an organic base which therefore exists as a side chain to the main nucleic acid chain (see figure 18.B).

Most remarkably, only two sugars and six bases occur in the structures of nucleic acids. The sugar ribose occurs in the RNA group of nucleic acids and deoxyribose occurs in the DNA group of nucleic acids. RNA stands for *r*ibose *n*ucleic *a*cid and DNA stands for *d*eoxyribose *n*ucleic *a*cid. RNA molecules are found as separate single chains but DNA consists of a pair of sugar-phosphate chains, intertwined and held together by hydrogen bonds between the side chain bases.

The function of nucleic acids is to act as templates for the production of all the varied proteins necessary for life. As many as two thousand different proteins are required in animals and their production is controlled by a 'code'. The codes are composed from only the six nucleic acid bases and their sequence along the sugar-phosphate chain.

For a fuller account of the polysaccharides and nucleic acids the Special Study *Biochemistry* should be consulted.

18.1 **Detergents**

Organic compounds which are not extracted by water but are extracted by non-polar solvents, such as ether, from plant or animal sources, are classified as lipids. Fats and waxes are the commonest materials classified as lipids and this section is concerned with fats and the derived detergent, soap.

Fats are esters of propane-1,2,3-triol, or glycerol

$$\begin{array}{l} CH_2OH \\ | \\ CHOH \\ | \\ CH_2OH \end{array}$$

with a variety of carboxylic acids, which have the general formula $CH_3(CH_2)_nCO_2H$ if saturated. Glycerol has three hydroxyl groups and can therefore combine with three carboxylic acid molecules, the commonest saturated acids having $n = 14$ or $n = 16$. Unsaturated acids also occur, the most common being $CH_3(CH_2)_7CH{=}CH(CH_2)_7CO_2H$ octadec-9-enoic (or oleic) acid. Because fats contain three ester links, owing to their glycerol content, they are also known as *triglycerides*, an example being the major component of castor oil. The molecular formula of this triglyceride is as now shown.

$$\begin{array}{rl} & \qquad\qquad\qquad\qquad\qquad\qquad\quad OH \\ & \qquad\qquad\qquad\qquad\qquad\qquad\quad | \\ & CH_2O_2C(CH_2)_7CH{=}CHCH_2CH(CH_2)_5CH_3 \\ & | \\ CH_3(CH_2)_5CHCH_2CH{=}CH(CH_2)_7CO_2 & CH \\ | \qquad\qquad\qquad\qquad & | \\ OH \qquad\qquad\qquad\qquad & CH_2O_2C(CH_2)_7CH{=}CHCH_2CH(CH_2)_5CH_3 \\ & \qquad\qquad\qquad\qquad\qquad\qquad\quad | \\ & \qquad\qquad\qquad\qquad\qquad\qquad\quad OH \end{array}$$

glucose

glucose polymerized to cellulose

glucose polymerized to starch

Figure 18.A
Polysaccharides.

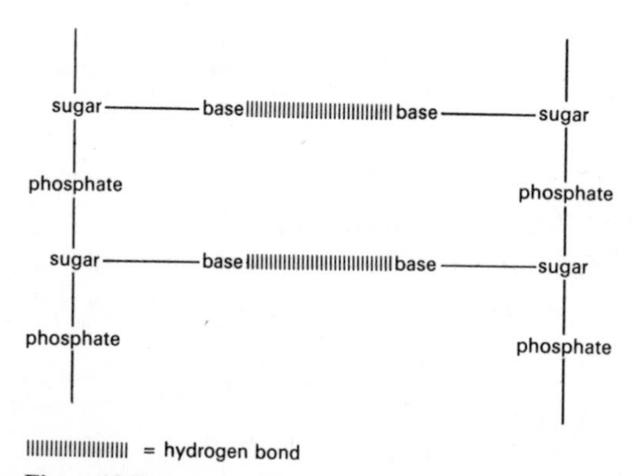

Figure 18.B
Part of a nucleic acid chain–DNA (schematic).

In this molecule, three molecules of ricinoleic acid form an ester with one molecule of glycerol.

The preparation of soap from fats and oils requires the hydrolysis of the ester linkages, using aqueous alkali. Modern soap making is a complex process but, on a small scale, the main reaction is easily performed.

Experiment 18.1a

Preparation of a soapy detergent

Castor oil is a convenient starting material for this experiment but it is not used industrially.

Dissolve 2 g of sodium hydroxide in 10 cm^3 of cold water (*caution*) and add 2 cm^3 of castor oil. Boil the mixture gently for about five minutes until the oily layer is no longer visible. Add water when necessary to maintain the volume at about 10 cm^3.

Dilute the boiled mixture with 10 cm^3 of water and saturate with sodium chloride (six spatula measures). Boil again for a minute, then cool and collect the solid product by suction filtration, washing free of alkali with a little water.

Make an aqueous solution of some of your product and examine its properties in the following test-tube reactions.

1 Add a little of your solution to pure water in a conical flask. Shake well. Is a lather formed?

2 Add a little of your solution to a calcium salt solution. Is a precipitate formed?

3 Add a little of your solution to dilute hydrochloric acid. Is a precipitate formed?

Is the behaviour of your product that of a soapy detergent? Is your product glycerol, ricinoleic acid, or sodium ricinoleate?

Experiment 18.1b

Preparation of Turkey Red oil, a soapless detergent

Turkey Red oil is made by treating castor oil with concentrated sulphuric acid below 35 °C. The main reaction is with the hydroxyl groups $\begin{matrix} \text{OH} \\ | \\ -\text{CH}- \end{matrix}$ of ricinoleic acid which are sulphated to give $\begin{matrix} \text{OSO}_3\text{H} \\ | \\ -\text{CH}- \end{matrix}$.

Add 2 cm^3 of concentrated sulphuric acid in small portions to 1 cm^3 of castor oil. Mix well by stirring with a glass rod and cool under the tap if the mixture becomes more than moderately warm. After a few minutes pour into an equal volume of water (*caution*: exothermic). The product, which is very impure, separates as an oil. Remove a small portion with a dropping tube and add to a large volume of pure water. Shake it well; is a lather produced?

Turkey Red oil was used as a wetting agent in the dyeing of fabrics. But the sulphation of natural oils is not a cheap way to make soapless detergents; it is better at the present time to use by-products from petroleum refining.

Synthetic soapless detergents

From the end of the nineteenth century, animal and vegetable fats and oils were needed more and more for food so a search was started for other detergents. A detergent is literally 'anything that washes clean'. Detergents, as the experiments have revealed, are compounds with an ionic functional group such as carboxyl or sulphate attached to a comparatively long hydrocarbon chain, as shown diagrammatically in figure 18.1a.

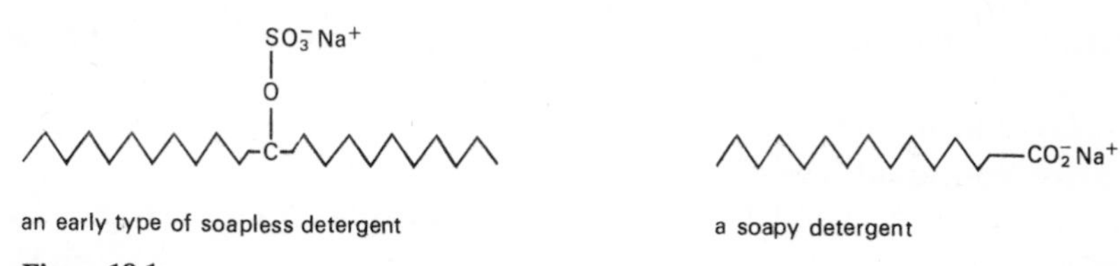

Figure 18.1a

What is required, ideally, are detergents that can be made from material not required for food, which do not form insoluble salts with calcium or magnesium ions, and which do not precipitate the free organic acids with mineral acid.

It has been noted that a detergent is formed when concentrated sulphuric acid is added to castor oil. The sodium alkyl sulphate so formed is a detergent and does not have some of the drawbacks of soap.

A wide range of by-products available from petroleum refining can be sulphated. The by-product selected for detergent manufacture should have the mean optimum length of hydrocarbon chain for good detergent properties: if the hydrocarbon chain is too short, the product will dissolve well in water but not in grease; if the hydrocarbon chain is too long, the product will dissolve well in grease but not in water.

Experiment 18.1c

To investigate the action of various sodium salts of carboxylic acids on the surface tension of water

In this experiment, the lowering of the surface tension of water by these substances is investigated. You are provided with 0.01M solutions of:

sodium acetate
sodium butanoate
sodium hexanoate
sodium octanoate
sodium decanoate
sodium dodecanoate

1 Write out the correct formulae of these salts.

Thoroughly wash and rinse seven test-tubes. Part fill six of the test-tubes with the 0.01M solutions provided and put distilled water in the seventh test-tube. Stand seven clean capillary tubes of uniform bore in the solutions and measure the height of the capillary rise in each case.

Plot a graph of capillary rise against number of carbon atoms in the salt.

2 Which salts lower the surface tension of water appreciably?
3 Which salt lowers it most?
4 What effect do you think sodium hexadecanoate (16 carbon atoms) will have on the surface tension of water?

Experiment 18.1d

Preparation of a synthetic soapless detergent

From the previous experiment it can be seen that dodecane derivatives should be successful starting materials for the preparation of detergents.

In this preparation dodecanol is sulphated with chlorosulphonic acid and the product neutralized with an organic base, triethanolamine, to form a liquid detergent.

$$C_{12}H_{25}OH + ClSO_3H \rightarrow C_{12}H_{25}OSO_3H + HCl$$
$$C_{12}H_{25}OSO_3H + N(CH_2CH_2OH)_3 \rightarrow C_{12}H_{25}OSO_3^- HN^+(CH_2CH_2OH)_3$$

Caution. Chlorosulphonic acid is more vigorous in its reactions than concentrated sulphuric acid. The wearing of protective gloves is therefore advised; also *all apparatus must by dry.*

In a fume cupboard clamp a 100 cm^3 beaker so that it can be cooled by raising a cold water bath around it. Place 7.5 cm^3 of dodecanol in the beaker and add 2.5 cm^3 of chlorosulphonic acid *in drops*. Hydrogen chloride gas will be evolved. Stir with a thermometer and ensure that the temperature remains at 35 ± 5 °C.

Allow the reaction mixture to stand for ten minutes and meanwhile prepare a solution of *either* 5 cm^3 of 2M sodium hydroxide in 50 cm^3 of water *or* 3 cm^3 of triethanolamine in 25 cm^3 of water.

To prepare a liquid detergent, add the reaction mixture in small portions to the triethanolamine solution. Any solid which forms should be rubbed into the liquid with a spatula until it dissolves. After the reaction mixture has all been added the complete mixture should be made neutral by the addition of sodium hydrogen carbonate in small portions.

To prepare a solid detergent, add the reaction mixture to the sodium hydroxide solution as above, making the complete mixture alkaline by addition of further small portions of sodium hydroxide solution. Evaporate to dryness on a water bath, testing occasionally to ensure the solution remains alkaline. The evaporation takes several hours.

Test the lathering ability of your synthetic soapless detergent against both soft and hard water.

How detergents work

The main problem in cleaning is the removal of grease. Dirt particles are usually held in place, for example on clothing, by a grease layer, so the removal of the grease will release most of the dirt particles. This is the basis of 'dry' cleaning processes in which clothing is agitated in organic solvents which dissolve grease.

How does the cleaning process work in the presence of water? Consider the soapy detergent molecule illustrated above. Which part of the molecule would mix with hydrocarbon grease and what intermolecular forces of attraction are involved? Which part of the molecule would mix with water and what intermolecular forces of attraction are involved?

Since the soapy detergent can mix with both water and grease the result is the removal of the grease from clothing to form an emulsion with the water. The process is illustrated in figure 18.1b. The dirt particles can similarly be held in suspension in the water by acquiring a surface layer of detergent molecules.

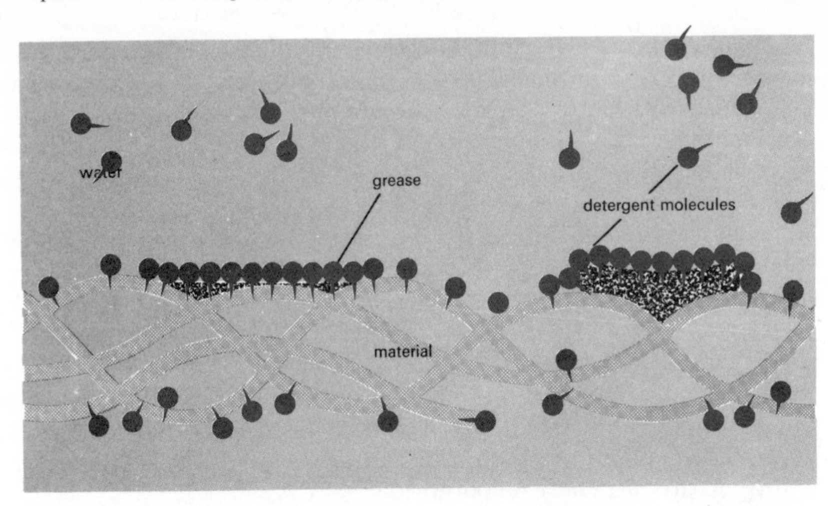

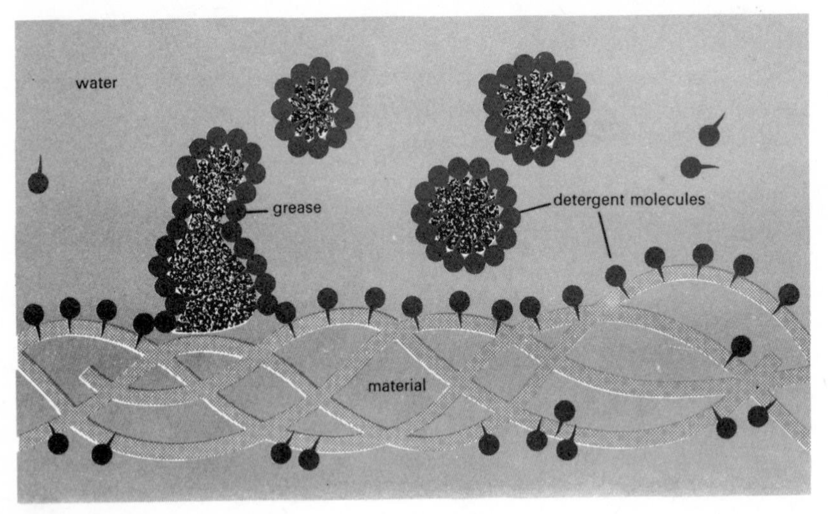

Figure 18.1b
The action of a detergent on grease.
Unilever

For the successful cleaning of clothes, detergent alone would be inadequate and commercial products contain a number of other essential ingredients.

A typical detergent powder for domestic use will contain:

20 *per cent active detergent* – As calcium salts of soapless detergents are soluble, calcium ions will not be removed from the water used for washing. If they are

not removed, they can migrate to the dirty fabric and form insoluble compounds with free fatty acids in the grease. They will also cause dirt particles that have been released to flocculate and redeposit on the fabric.

30 *per cent complex phosphates* – These are therefore added to form soluble complexes with the calcium ions and thereby inactivate them.

2 *per cent sodium carboxymethyl cellulose* – This is absorbed on both the dirt particles and the material being washed increasing the negative charges so that the suspended dirt is more strongly repelled from the material.

10 *per cent soluble silicates such as sodium or potassium* – These are added to prevent corrosion of some metals such as aluminium and they help to produce a crisp powder that is easy to handle.

5 *per cent sodium perborate* – This reacts with water forming hydrogen peroxide which acts as a mild bleach for chemical stains such as tea or fruit juices.

20 *per cent sodium sulphate* – This helps to make powder which is easy to handle.

The product also contains small quantities of foam stabilizers, perfume, colouring, and *fluorescers*. Fluorescers are materials which absorb ultraviolet light and re-emit the absorbed energy as visible blue light, thus counteracting the gradual yellowing of white materials.

The variety of synthetic soapless detergents

The first successful synthetic soapless detergents were based on the sulphation of alcohols as in the experiment above. The products are known as sodium alkyl sulphates.

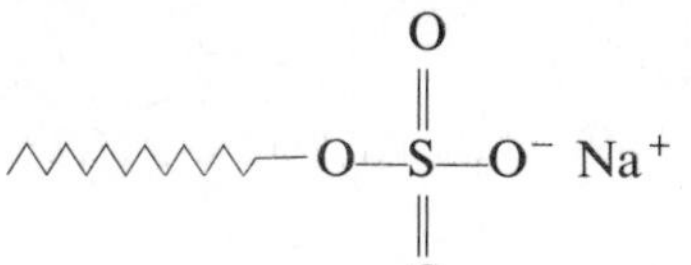

$$\text{R—O—}\overset{\text{O}}{\underset{\text{O}}{\overset{\|}{\underset{\|}{\text{S}}}}}\text{—O}^{-}\ \text{Na}^{+}$$

a sodium alkyl sulphate

After the Second World War the manufacture of other detergents was started.

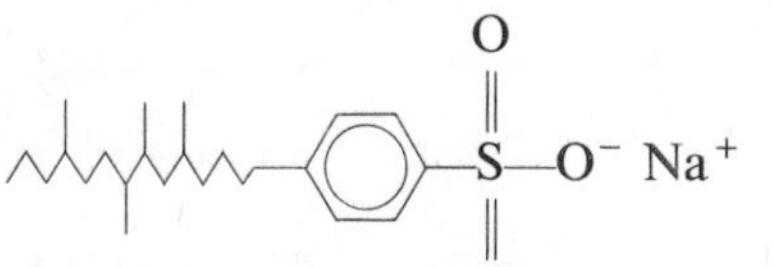

$$\text{R—C}_6\text{H}_4\text{—}\overset{\text{O}}{\underset{\text{O}}{\overset{\|}{\underset{\|}{\text{S}}}}}\text{—O}^{-}\ \text{Na}^{+}$$

a sodium alkyl aryl sulphonate
(note the side chains)

At this time benzene derivatives became more easily obtainable, and it is easier to replace a hydrogen atom in the benzene ring by a *sulphonate* group ($—SO_3^-$) than a hydrogen atom in an alcohol by a sulphate group ($—O—SO_3^-$). Both processes use concentrated sulphuric acid. Petroleum companies can supply large quantities of alkyl benzene for sulphonation:

However the use of such compounds gave rise to serious problems at sewage works and on rivers whose proper functioning was upset by persistent foam (figure 18.1c). The remedy has been to change to compounds with straight-chain alkyl groups. These can be degraded by bacteria and thus lose their detergent properties

O
‖
—S—O^- Na^+ a biodegradable detergent
‖ (note the absence of side chains)
O

Figure 18.1c
One result of the use of detergents with a branched chain alkyl group (Batford Weir on the River Lee). Taken in 1961.
Photo, Crown Copyright

95 per cent of soapless detergents sold at present are similar to this one, that is, the head carries a negative charge.

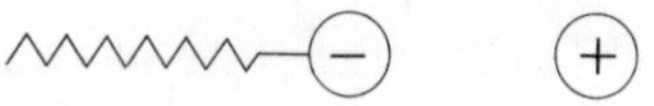

Detergents are also made in which the detergent molecule is a cation:

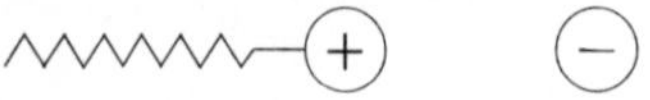

such as

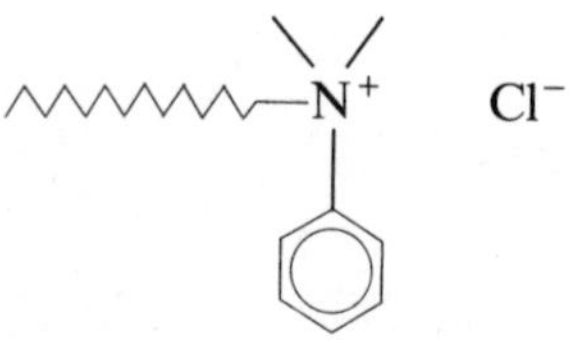

Cl^-

These are not good washing agents unless concentrated, but they have disinfectant properties.

Non-ionic detergents are also made and their importance is growing. They are particularly useful in liquid soapless detergent products. They are useful because they remain active in the presence of acid, alkali, and even high concentrations of electrolytes. They do not foam to any extent but foam is not necessary for cleaning action in washing. Foam has a psychological significance as a pointer to cleaning action, but this significance is diminishing with the growth of machine washing.

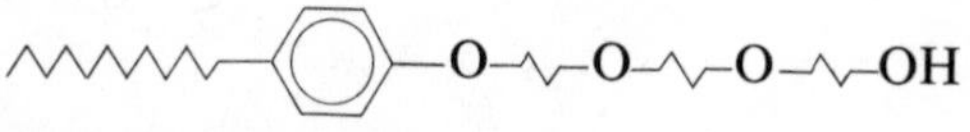

—O—O—O—OH

Non-ionic detergents can be used to help disperse oil slicks at sea. When a ship is wrecked, very large quantities of oil may be released and its dispersal is difficult. Unfortunately some detergents are more toxic to marine life than the oil itself and the process of cleaning a rocky foreshore can kill the marine life which made the foreshore an interesting and attractive place (figure 18.1d).

Figure 18.1d
i Oil on Prah Sands, Cornwall, 27th March 1967, from the wrecked oil tanker *Torrey Canyon*. *Photo*, *Keystone*

ii Limpets at Trevone, Cornwall, 15th April 1967, a day or two after detergent spraying nearby. Two limpets have lost their shells leaving their soft parts in situ. *Photo, D. P. Wilson*

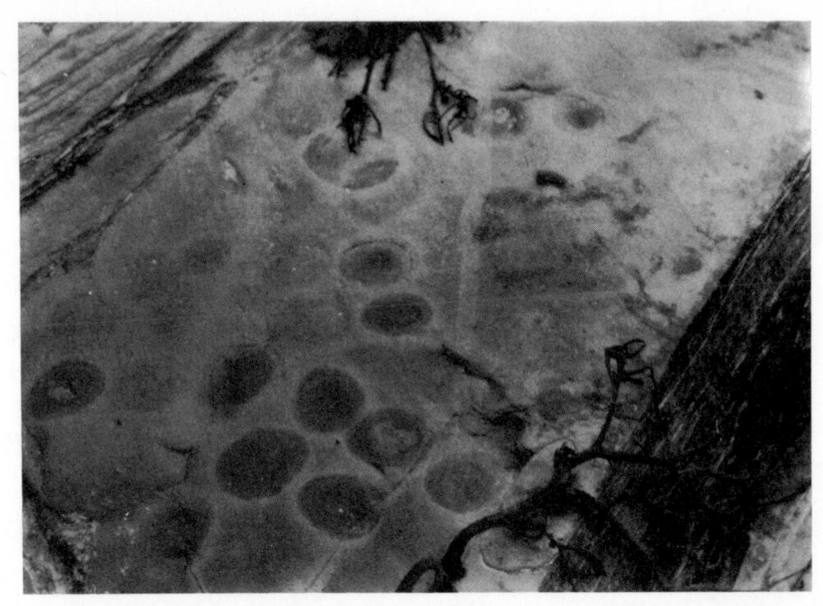

iii The same rock on 23rd April 1967. The dead limpets have been washed away. *Photo, D. P. Wilson*

18.2 **The nature of proteins**

Chemical tests reveal the presence of nitrogen compounds in most animal tissues and, more specifically, the acidic hydrolysis of hair, blood, and muscle tissue shows that they consist almost entirely of amino acids. Naturally occurring compounds made from amino acids are known as proteins.

$$\underset{\displaystyle\mathrm{R}}{\mathrm{NH_2{-}\underset{|}{CH}{-}CO_2H}} \qquad \text{an amino acid}$$

$$\mathrm{{-}NH{-}\underset{\underset{\displaystyle R}{|}}{CH}{-}CO{-}NH{-}\underset{\underset{\displaystyle R}{|}}{CH}{-}CO{-}NH{-}\underset{\underset{\displaystyle R}{|}}{CH}{-}CO{-}} \qquad \text{part of a protein}$$

The term *peptide group* is used to describe the amide group —CO—NH— when it links together amino acid residues.

Write out the equation for the formation of a peptide group between two amino acids. The organic product is known as a dipeptide; what is the other product of the reaction and what type of reaction has occurred?

When only a few amino acid residues are connected by peptide groups the molecules are known as *peptides*, but when more than fifty amino acid residues are involved the term *protein* is used.

Experiment 18.2

Investigation of protein materials

In principle every foodstuff and all materials of biological origin are worth testing for protein content. However, as the tests suggested are carried out with solutions it would be as well to select water soluble substances.

Fresh milk, egg (white and yolk) or the derived extracts casein and albumen can be tested. Other possibilities are pepsin and trypsin, which are digestive enzymes; gelatin, from the hydrolysis of the connective tissue of animals; and commercial meat extracts. Glycine, as a simple amino acid, and casein hydrolysate (containing the free amino acids of milk), might also be tried.

Dissolve a small sample of material in water, warming if necessary, divide into two portions, and test as follows:

1 Add an equal volume of 2M sodium hydroxide to your cool solution followed by 1 *drop* of 0.1M copper(II) sulphate. A mauve colour will develop if proteins are present. This is known as the *biuret test* and detects peptide groups:

$$
\begin{array}{ccccccccc}
 & O & & & & O & & & \\
 & \| & & & & \| & & & \\
— & C & — & NH & — CH — & C & — & NH & — \\
 & & & & | & & & & \\
 & & & & R & & & &
\end{array}
$$

What type of reaction and bonding would you expect between copper ions and peptide groups?

Carry out a blank test using water as your 'protein sample'.

2 Add a few drops of 0.02M ninhydrin solution to your solution and boil gently for a minute. A red to blue colour will develop if proteins or amino acids are present. The *ninhydrin test* detects free $—NH_2$ groups and it is an excellent colour reaction for amino acids. Details of the reaction (which you are not expected to learn) are given in figure 18.2a.

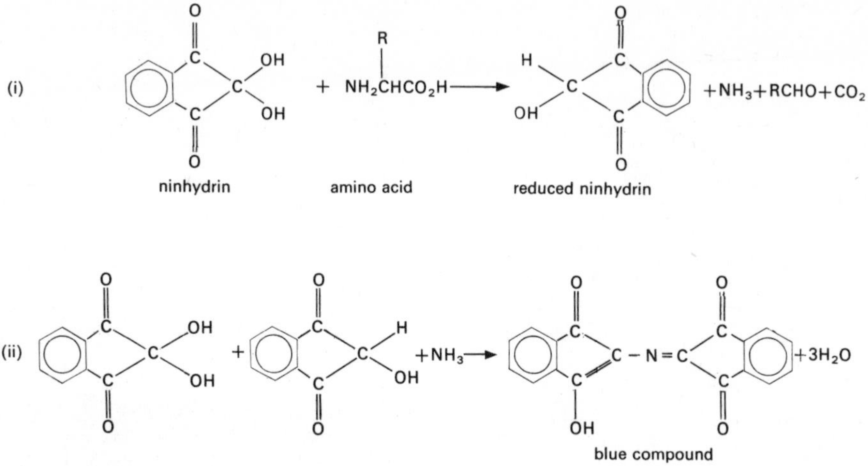

(i) ninhydrin + NH_2CHRCO_2H (amino acid) $\longrightarrow$ reduced ninhydrin $+ NH_3 + RCHO + CO_2$

(ii) ninhydrin + reduced ninhydrin $+ NH_3 \longrightarrow$ blue compound $+ 3H_2O$

Figure 18.2a
Development of Ninhydrin Blue.

3 If you have enough time, spot amino acid solutions on to filter paper using a capillary tube, spray with ninhydrin solution, and heat in an oven at 110 °C until any coloured spots are well developed. Make a note of any unexpected coloured areas. How long a heating period was necessary? Record your results in the form of a table, noting the particular colour tones obtained.

The biological importance of proteins is apparent from their widespread occurrence and variety of function (table 18.2a and figure 18.2b).

Equally striking is the complexity of proteins, with molecular weights of a thousand or more; but at the same time proteins illustrate how in nature complex ends are often achieved through the infinitely varied use of simple means. All the different naturally occurring proteins are built not as might be expected from many hundreds of different amino acids but from only two dozen.

The amino acids found in naturally occurring proteins are all α-amino carboxylic acids and further they are all L-amino acids. In α-amino acids the amino group occurs on the carbon atom adjacent to the acid group. The stereochemistry of amino acids was discussed in section 13.5 and you should refer again to that section to revise your understanding of D and L configurations.

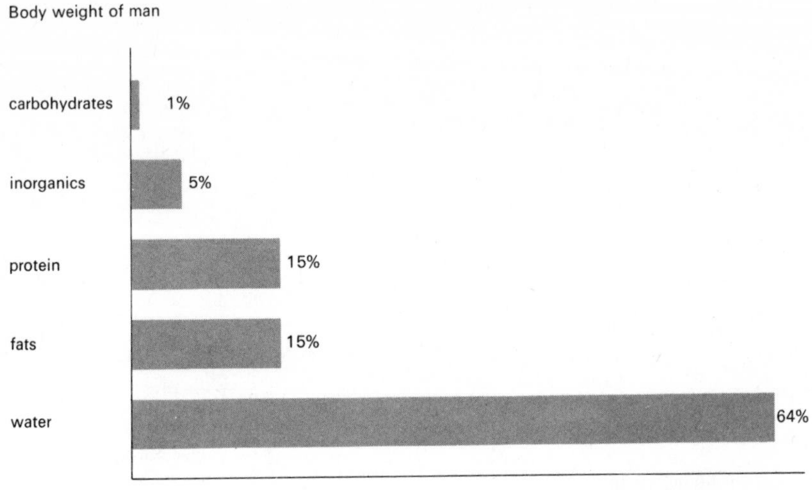

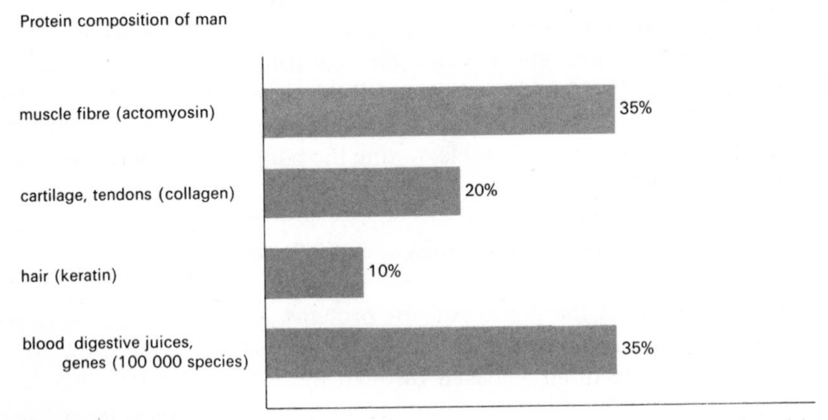

Figure 18.2b
The importance of proteins in man.

Some typical amino acids are listed in table 18.2b which you should now examine. Since there are many proteins and few amino acids in nature it is apparent that proteins will be characterized by the sequence in which their amino acids are linked:

—gly—lys—glu—
or —lys—gly—glu—
or —lys—glu—gly—etc.

The determination of the amino acid sequence of a protein is considered in the next section.

Table 18.2a
The occurrence and function of some proteins

Protein	Occurrence (examples)	Function	Molecular weight	Approximate number of amino acid units
Insulin	animal pancreas	governs sugar metabolism	5700	51
Ribonuclease	animal and plant cells	breaks down ribonucleic acid	13 700	124
Lysozyme	white of egg	dissolves cell wall of bacteria	14 400	119
Myoglobin	muscle	oxygen carrier	17 000	153
Trypsin	animal pancreas	digests food proteins	23 800	180
Haemoglobin	blood	oxygen carrier	68 000	574
Edestin	hemp seed	?	320 000	?
Urease	soya beans	converts urea to ammonia	480 000	4500

Table 18.2b
Some typical amino acids

Formula	Name	Abbreviation	Nature of side chain	R_f value in butan-1-ol/acetic acid/water (5/1/4)
H_2NCHCO_2H with side chain H	glycine	gly	neutral	0.26
H_2NCHCO_2H with side chain CH_3	alanine	ala	neutral	0.38
H_2NCHCO_2H with side chain $CHCH_3$, CH_3	valine	val	neutral	0.60

Formula	Name	Abbreviation	Nature of side chain	R_f value in butan-l-ol/acetic acid/water (5/1/4)
H_2NCHCO_2H | CH_2 | $CH(CH_3)_2$	leucine	leu	neutral	0.73
H_2NCHCO_2H | CHC_2H_5 | CH_3	isoleucine	ile	neutral	0.72
H_2NCHCO_2H | CH_2OH	serine	ser	neutral	0.27
H_2NCHCO_2H | CHOH | CH_3	threonine	thr	neutral	0.35
H_2NCHCO_2H | CH_2CO_2H	aspartic acid	asp	acidic	0.24
H_2NCHCO_2H | CH_2 | CH_2CO_2H	glutamic acid	glu	acidic	0.30
H_2NCHCO_2H | CH_2SH	cysteine	cys	acidic	0.08

Formula	Name	Abbreviation	Nature of side chain	R_f value in butan-1-ol/acetic acid/water (5/1/4)
H_2NCHCO_2H — CH_2S — CH_2S — H_2NCHCO_2H (S–S linked)	cystine	cys-cys	neutral	—
H_2NCHCO_2H — CH_2 — CH_2SCH_3	methionine	met	neutral	0.55
H_2NCHCO_2H — $(CH_2)_3$ — NH — $HN{=}C{-}NH_2$	arginine	arg	basic	0.16
H_2NCHCO_2H — $(CH_2)_3$ — CH_2NH_2	lysine	lys	basic	0.14
H_2NCHCO_2H — $CH_2{-}C_6H_5$	phenylalanine	phe	neutral	0.68
H_2NCHCO_2H — $CH_2{-}C_6H_4{-}OH$	tyrosine	tyr	acidic	0.50

18.3 The chemical investigation of proteins

To establish the correct molecular formula of a substance it must be available pure and the necessary experimental techniques must be available. F. Sanger, working at Cambridge, was the first chemist to establish the molecular formula of a protein. When he started work in 1944, he had to develop new experimental techniques because the problems of protein composition were unsolved at that time and the only simple protein available pure was insulin. Ten years' work was necessary to establish the correct amino acid sequence for insulin.

The molecular formula of insulin is now known to be $C_{254}H_{377}N_{65}O_{75}S_6$! As is the case with any protein, the first stage in the investigation of insulin was to discover the nature and number of amino acid residues present.

Hydrolysis of a protein by refluxing with 6M hydrochloric acid for 24 hours produces the free amino acids. (Why should a hydrolysis reaction break the peptide bonds of the protein?) Quantitative separation of the hydrolysate will determine which amino acids are present and their relative amounts. Sanger found 17 different amino acids in insulin ranging from six units of cysteine and leucine to one unit of lysine, the molecular formula being accounted for by a total of 51 amino acid units.

The separation of the amino acids is achieved by paper chromatography. The method depends on the 'partition' of the amino acids between two different solvents (as in an ether extraction). One of the solvents is water which is bonded to the paper and therefore stationary ('dry' cellulose can contain by weight 2 per cent hydrogen bonded and 5 per cent capillary water). The other solvent, which is allowed to move over the paper, is chosen to emphasize solubility differences among the amino acids. Thus if a butan-1-ol/acetic acid mixture is chosen, an amino acid such as valine is more soluble in the mixture than in water, and therefore is carried over the paper by the mixture. On the other hand, cysteine is less soluble in the mixture than in water, and therefore remains stationary on the paper.

For the complete chromatographic separation of amino acids it is necessary to use first one solvent and then a second different solvent which is allowed to move over the paper at right-angles to the direction of the first solvent. Examine the two-way chromatogram given in figure 18.3b and mark the amino acids known to you as acidic, basic, or neutral; what chemical nature would you predict for the other amino acids? Which amino acids are relatively more soluble in water than in the phenol solvent system?

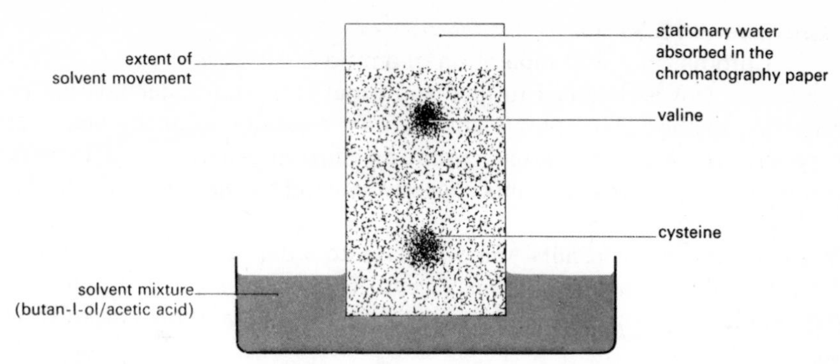

Figure 18.3a
Representation of paper chromatography of amino acids.

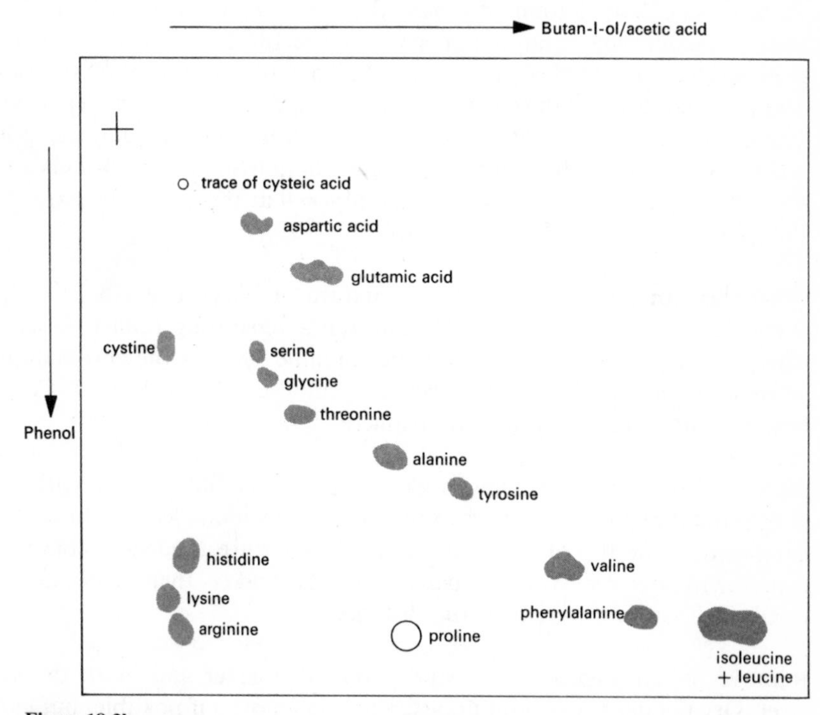

Figure 18.3b
Paper chromatography separates the 17 amino acids of insulin. In the chromatogram represented by this diagram insulin was broken down by hydrolysis and a sample of the mixture placed on the upper left corner of the sheet of paper. The sheet was hung from a trough filled with the solvent which carried each amino acid a characteristic distance down the paper. The sheet was then turned through 90 degrees and the process repeated. The amino acids, with the exception of proline, appear as purple spots when sprayed with ninhydrin. *After Thompson, E. O. P. (1955) 'The insulin molecule'.* Scient. Am. **192**, *5. Copyright © 1955, by* Scientific American, *Inc. All rights reserved.*

Experiment 18.3a

Chromatographic separation of amino acids

The experiment is designed to give you experience and understanding of an important method. To separate and identify naturally occurring amino acids by paper chromatography would require an effort spread over about three days, so this brief experiment can only suggest the potentialities of the method.

To obtain satisfying results you will have to work with care and keep the experimental materials scrupulously clean. Touch the chromatography papers only on their top corners and never lay them down except on a clean sheet of paper.

Procedure

Place spots of 0.01M amino acids in aqueous solution 1.5 cm from the bottom edge of the chromatography paper (see figure 18.3c, in which a CRL/1 paper is illustrated). To do this, dip a *clean* capillary tube in the stock solution and apply a small drop to the chromatography paper using a quick delicate touch. Practise on a piece of ordinary filter paper until you can produce spots not more than 0.5 cm in diameter. Apply spots of individual amino acids and also mixtures, making identification marks in *pencil* at the top of the paper. Allow the spots to dry.

Meanwhile prepare a fresh solvent mixture of butan-1-ol (15 cm^3), glacial acetic acid (3 cm^3), and water (12 cm^3) in a separating funnel. Discard the lower aqueous layer and use the upper organic layer as the chromatography solvent by running it into a covered 1 cubic decimetre beaker. Cover the beaker to produce a saturated atmosphere.

Now roll the chromatography paper into a cylinder and secure it with a paper clip. Stand the cylinder in the covered solvent beaker and leave it for the solvent to ascend to the top of the slots. As 20 minutes are needed to complete the experiment after removing the paper from the beaker, there may not be time to allow the solvent to rise the full distance.

Remove the chromatography paper from the beaker and mark the solvent level. Dry the paper (without unfastening), in an oven if possible, but *not* over a Bunsen flame, because the solvent is both pungent and inflammable.

Detect the amino acids by spraying the paper sparingly with 0.02M ninhydrin solution and then heating in an oven at 110 °C for 10 minutes. Purple spots should appear.

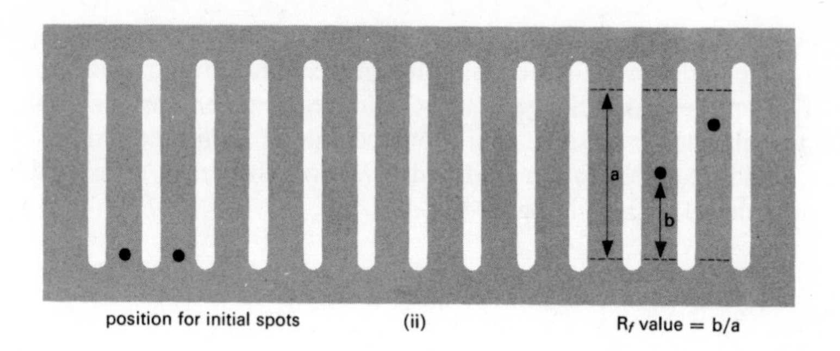

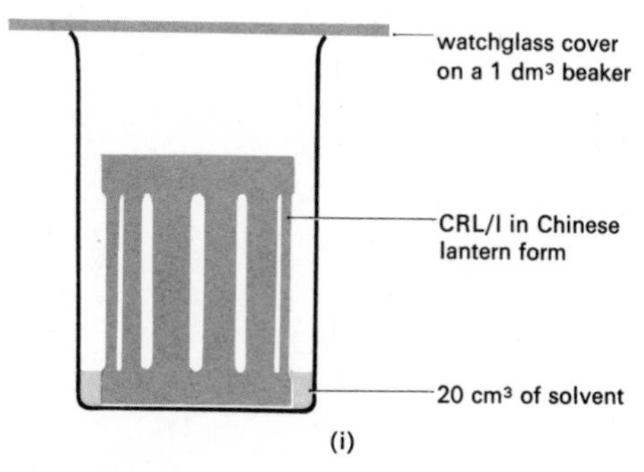

Figure 18.3c
Apparatus for simplified chromatography of amino acids.
After Rendle, G. P., Vokins, M. D. W., and Davis, P. H. H. (1967) Experimental chemistry, *Edward Arnold*

Preserve the spots by spraying with a mixture made up of methanol (19 cm^3), M aqueous copper(II) nitrate (1 cm^3), and 2M nitric acid (a drop), and then expose *in a fume cupboard* to the fumes from 0.880 ammonia. Determine the R_f value of amino acid samples.

R_f values are obtained using the expression

$$R_f \text{ value} = \frac{\text{distance moved by amino acid}}{\text{distance moved by solvent}}$$

Have the mixtures separated?

After the chromatographic identification of the amino acids present in a protein the next stage in the investigation is to determine which amino acids are on the ends of the peptide chains. Sanger found a reactive benzene compound (1-fluoro-2,4-dinitrobenzene or FDNB) that would combine with free amino groups giving a bonding stable to the acid hydrolysis which breaks up peptide. The reaction sequence is given in figure 18.3d.

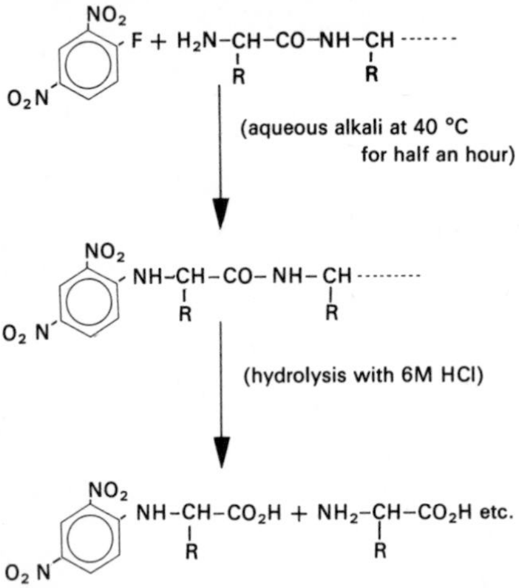

$$2,4\text{-}(O_2N)_2C_6H_3\text{-}F + H_2N\text{-}\underset{R}{CH}\text{-}CO\text{-}NH\text{-}\underset{R}{CH}\cdots$$

(aqueous alkali at 40 °C for half an hour)

$$2,4\text{-}(O_2N)_2C_6H_3\text{-}NH\text{-}\underset{R}{CH}\text{-}CO\text{-}NH\text{-}\underset{R}{CH}\cdots$$

(hydrolysis with 6M HCl)

$$2,4\text{-}(O_2N)_2C_6H_3\text{-}NH\text{-}\underset{R}{CH}\text{-}CO_2H + NH_2\text{-}\underset{R}{CH}\text{-}CO_2H \text{ etc.}$$

Figure 18.3d
The reaction of FDNB with the terminal amino group of a protein.

The N-terminal amino acid is thereby labelled and can be recovered as a yellow dinitrophenyl amino acid. Chromatography is used to identify the dinitrophenyl amino acid. Experiment 18.3b illustrates the procedure.

Experiment 18.3b

The N-terminal analysis of a dipeptide

By determining which of the two amino acids in a dipeptide has a free amino group it is possible to establish the amino acid sequences

either $H_2N\text{—}\underset{R_1}{CH}\text{—}CO\text{—}NH\text{—}\underset{R_2}{CH}\text{—}CO_2H$

or $H_2N\text{—}\underset{R_2}{CH}\text{—}CO\text{—}NH\text{—}\underset{R_1}{CH}\text{—}CO_2H$

This process is referred to as N-terminal analysis.

In this method 1-fluoro-2,4-dinitrobenzene (FDNB) is reacted under mildly alkaline conditions with the free amino group of the peptide, to yield a dinitrophenyl (DNP) peptide. After acid hydrolysis to split up the peptide group the yellow DNP amino acid can be identified by paper chromatography.

The equations of the reactions are given in figure 18.3d.

Warning. FDNB can blister the skin and a pair of disposable plastic gloves must be worn during the experiment. All apparatus should be washed with ammonia after use to make any FDNB innocuous.

Dissolve 10 mg (*no more*) of a dipeptide in 2 cm^3 of 0.4M sodium hydrogen carbonate in a flask which can be stoppered (a 50 cm^3 glass-stoppered pear shaped flask) and add 1 cm^3 of a 2 per cent (v/v) solution of FDNB in ethanol. This is an approximately threefold excess of FDNB. Now heat the flask in a water bath at 40 °C for 30 minutes.

Allow it to cool completely then extract twice with ether (3–5 cm^3 portions) to remove any excess of FDNB. Add the ether directly to the flask, shake it well, and remove the upper layer of ether using a dropping tube. The ether extracts should be treated with ammonia before being discarded.

By means of a dropping tube transfer the lower aqueous layer to a clean flask, leaving behind any residual ether, and add 2 cm^3 of concentrated hydrochloric acid. Heat in a boiling water bath for 30 minutes to hydrolyse the peptide group.

Allow to cool completely then extract twice with ether (3–5 cm^3 portions) as before. The DNP amino acid will be extracted by the ether and is obtained by evaporating the combined ether extracts on a hot water bath. (*Caution.* Turn out all Bunsen burners.)

For chromatography the yellow DNP amino acid is dissolved in 2 cm^3 of acetone and spotted onto a CRL/1 paper. Great care is necessary to avoid overloading the paper, otherwise considerable 'tailing' of the spots will occur during development. If capillary melting point tubes are used for spotting they should be drawn out to half their diameter. On CRL/1 paper a number of separate spots can be placed, diluting the DNP amino acid with more acetone between each application.

Shake butan-1-ol and water together in a separating funnel and use 20 cm^3 of the upper layer (water-saturated butan-1-ol) as solvent to develop the chromatogram.

The chromatogram requires about 80 minutes for full development on CRL/1 paper (a 9 cm rise) and, as DNP amino acids are light-sensitive, the apparatus is best placed in a dark cupboard.

After development, mark the solvent height and then dry the chromatogram briefly in a warm oven. Yellow spots may be seen due to 2,4-dinitroaniline (R_f 0.95) and 2,4-dinitrophenol (R_f 0.55–0.60). These spots can be faded by holding the chromatogram in hydrogen chloride fumes (a little ammonium chloride plus concentrated sulphuric acid in an evaporating basin). The remaining yellow spot will be due to either DNP leucine (R_f 0.75) or DNP glycine (R_f 0.30–0.35).

Determine by measurement of R_f value (distance moved by DNP amino acid/distance moved by solvent), which DNP amino acid is present and hence whether your original dipeptide was glycyl-leucine or leucyl-glycine.

What happened to the other amino acid?

Sanger's N-terminal analysis of insulin revealed two terminal amino groups per molecule, the associated amino acids being glycine and phenylalanine.

As Sanger found two terminal amino groups, how many chains should the insulin molecule consist of? Are there any amino acids capable of forming a bridge between the chains? Make a drawing, in outline only, of a possible structure.

To determine the detailed structure of a protein the above techniques are elaborated: partial hydrolysis to di- and tri-peptides followed by separation and investigation by the dinitrophenyl method described above will provide further evidence of the amino acid sequence.

The use of enzymes can provide a specially delicate method of hydrolysis. Thus the enzyme trypsin will break up only peptide groups involving the carboxyl group of basic amino acids such as lysine or arginine. Other enzymes can be used but they are less selective, for example chymotrypsin breaks protein chains mainly at the carboxyl group of amino acids with aromatic side chains.

Thus the use of enzymes provides polypeptides of moderate chain length and thereby simplifies the investigation of amino acid sequence. Examine the amino acid sequences given in figure 18.3e, and mark the peptide groups that will be hydrolysed by trypsin.

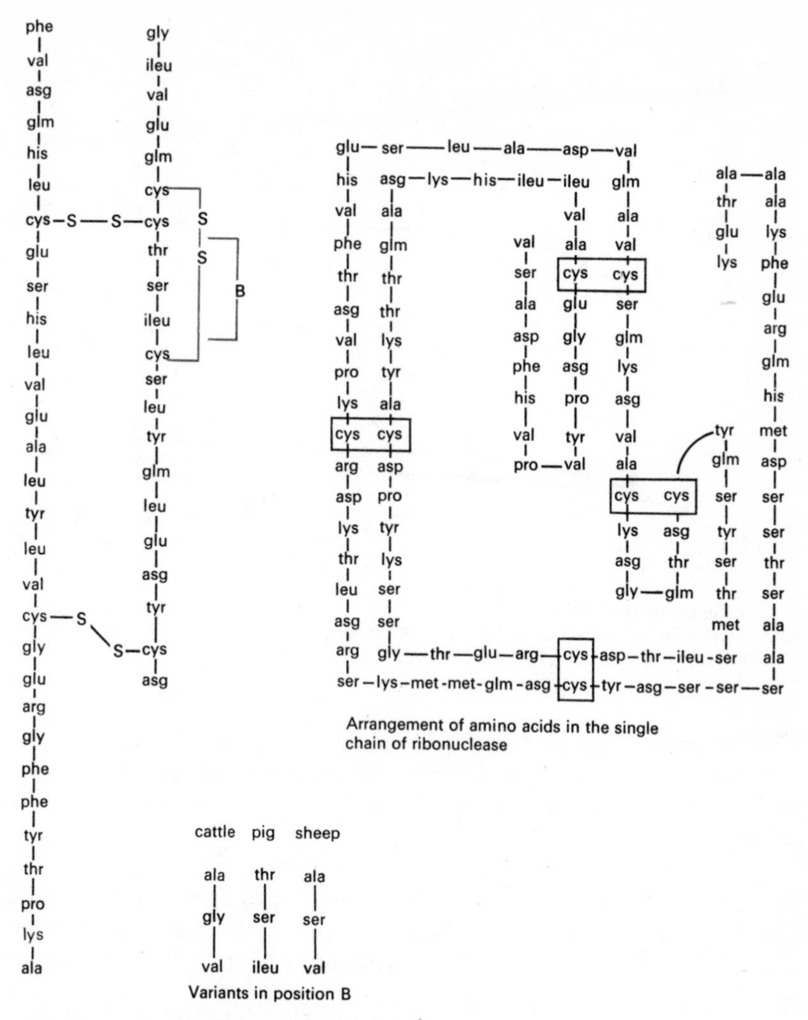

The arrangement of the fifty-one amino acids in the molecule of insulin. Insulins from other species differ slightly in the short section marked B

Figure 18.3e
The amino acid sequence of insulin and ribonuclease.

Another technique of interest is the use of an enzyme, carboxypeptidase, which breaks peptide bonds in sequence starting at the free carboxyl end of protein chains. A typical result is illustrated in figure 18.3f.

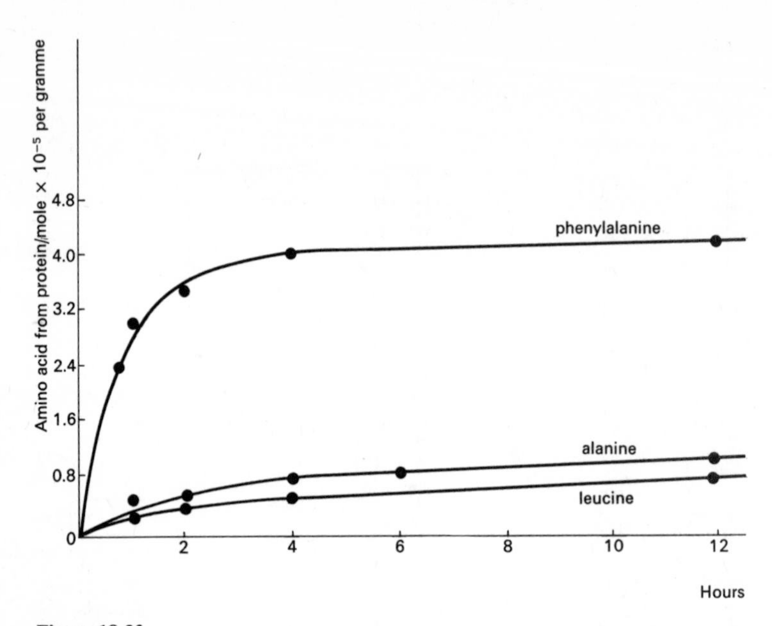

Figure 18.3f
Rate of release of amino acids by the action of carboxypeptidase on somatropin (molecular weight 45 000). Enzyme:substrate – 1:50; pH 8.5 at 25 °C.
After Harris, Li, Condliffe, and Penn (1954) J. Biol. Chem. **133**, *209*

What interpretation can you make of the graph, which shows the variation with time of free amino acid concentration from the hydrolysis of a protein by carboxypeptidase? Calculate the number of moles of amino acid produced per mole of protein.

By the use of the techniques described a number of amino acid sequences have now been determined, the complexity of the work leading to automation of some of the separation procedures. Typical results are indicated in table 18.3 and figure 18.3e.

Protein	Number of amino acid units	Year in which the amino acid sequence was discovered
Insulin	51	1954
Ribonuclease	124	1960
Tobacco mosaic virus	158	1961
Haemoglobin	574	1963
Myoglobin	151	1964

Table 18.3
The discovery of the amino acid sequences in proteins

To appreciate the jigsaw puzzle nature of the work you can attempt the following exercise:

Set out the sequence L U O C S O N four times to represent polypeptide molecules and cut each one up randomly into singles and pairs. Now use each set of pieces to deduce the original sequence. Is there only one possible solution?

Now repeat the exercise with the sequence D I O T N O C A O R. Is there still only one solution possible from the evidence? If more than one, how many?

Alternatively, produce your own sequences of seven and ten letters and provide your neighbour with the cut pieces to see if he can deduce your original sequence. Note that the seven letter sequence has one letter repeated twice and the ten letter sequence has one letter repeated three times.

When the amino acid sequence of a protein has been established the next problem is to establish the three-dimensional shape adopted by the polypeptide chain. The solving of this problem is considered in the next section.

18.4 The structural investigation of proteins

There are numerous possibilities for the shape of a molecule of insulin. Rotation about the single bonds in the polypeptide chain makes possible the different shapes, which are known as *conformations*. It is even possible that in different situations the protein molecule will adopt different conformations; however if a protein can be obtained in a crystalline state this is good evidence that all the molecules have the same conformation in the solid state. And the protein crystals can be used for X-ray diffraction studies.

The peptide group

What is the chain length of the insulin molecule (with 51 amino acids)? Assume the bonds in the peptide group have an average length of 0.14 nanometres and cut a piece of string to represent the insulin chain on the scale of 0.1 nm = 1 cm. Imagine how many different ways there are of coiling up your piece of string!

Fortunately a polypeptide chain is not as flexible as a piece of string and the first requirement in investigating the possible conformations of a protein is to know as much as possible about the shape and properties of the peptide group. To this end X-ray diffraction studies have been made of simple compounds such as acetylglycine:

$CH_3CONHCH_2CO_2H$
acetylglycine

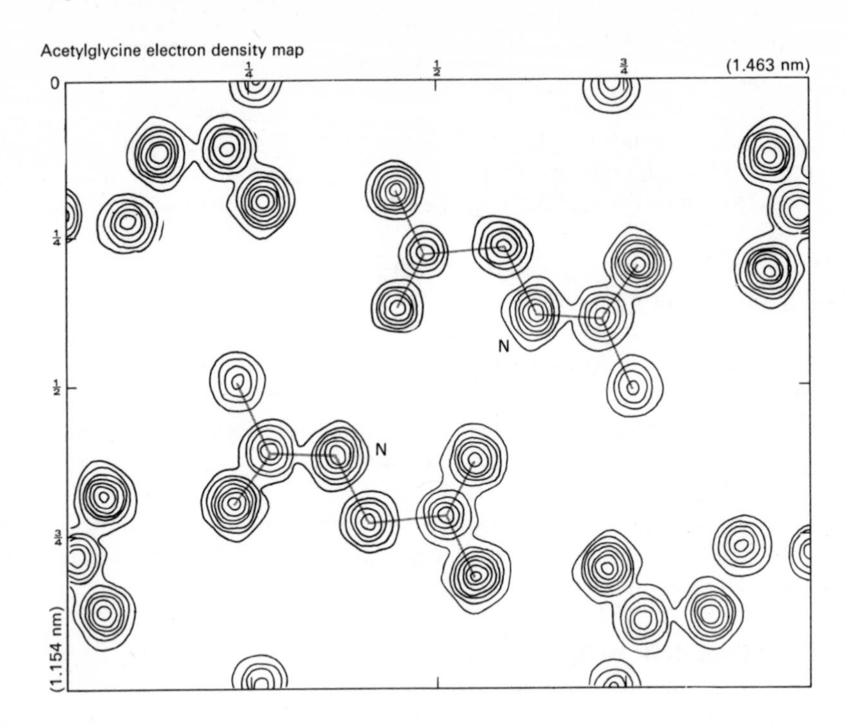

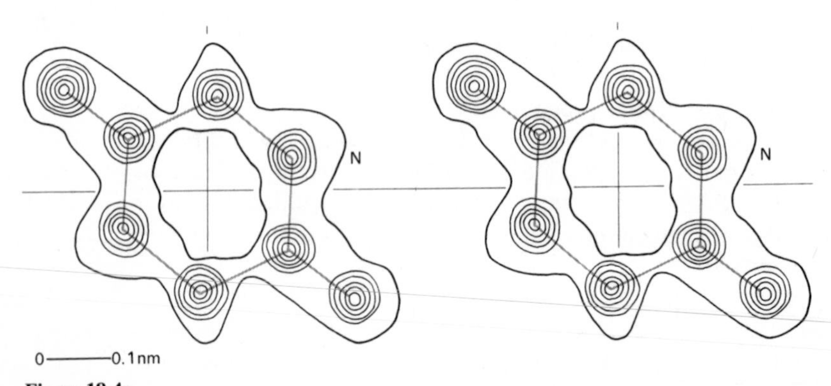

Figure 18.4a
Electron density maps.
(*top*) *After Carpenter, G. B. and Donohue, J.* (*1950*) J.A.C.S. **72**, *2318*
(*bottom*) *After Dageilh, R. and Marsh, R. E.* (*1959*) Acta Cryst. **12**, *1007*

Hydrogen bonding must also be considered; for example, the polypeptide chains of silk are held together by hydrogen bonds (Topic 10, figure 10.2h). The shape of the hydrogen bond in proteins is best investigated by X-ray diffraction studies on simple compounds like the cyclic anhydride of glycine (diketopiperazine). The relationship of the molecules in the crystal suggests that they are held together by hydrogen bonds. The appropriate intermolecular distances will thus be a measure of hydrogen bond length in proteins.

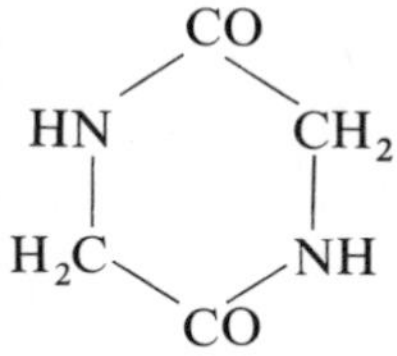

```
        CO
      /    \
  HN        CH2
   |         |
  H2C       NH
      \    /
        CO
```

diketopiperazine

Experiment 18.4a

The structure of the peptide group

Electron density maps of acetylglycine and the cyclic anhydride of glycine are given in figure 18.4a.

Examine these electron density maps, relate them to the formulae of the compounds, and identify the atoms. Remember that hydrogen atoms are not revealed by X-rays and that, as oxygen has more electrons than carbon, it is likely to be characterized by more contour lines on the density map. Measure the bond lengths and angles in the peptide group of acetylglycine:

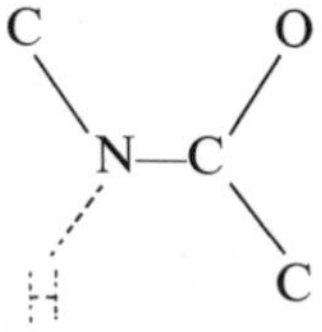

```
  C         O
   \       /
    N — C
   ⋰      \
  H         C
```

A more detailed examination of the evidence suggests that the six atoms involved in the peptide group are all in one plane.

What bonds are necessary to hold all the atoms in one plane (co-planar)? Is this the conventionally expected bonding? From your bond lengths calculated from the X-ray diffraction patterns, which are the single bonds and which have double-bond character?

Usual single-bond lengths are C—C, 0.154 nm; C—O, 0.143 nm; C—N, 0.147 nm.

Measure the hydrogen bond length O┆┆┆┆┆┆┆H—N between molecules of diketopiperazine. Would you expect a bond of this magnitude of length to be strong or weak?

By giving due weight to all the experimental evidence from the study of a number of dipeptides, an average peptide group can be described. The data are given in figure 18.4b.

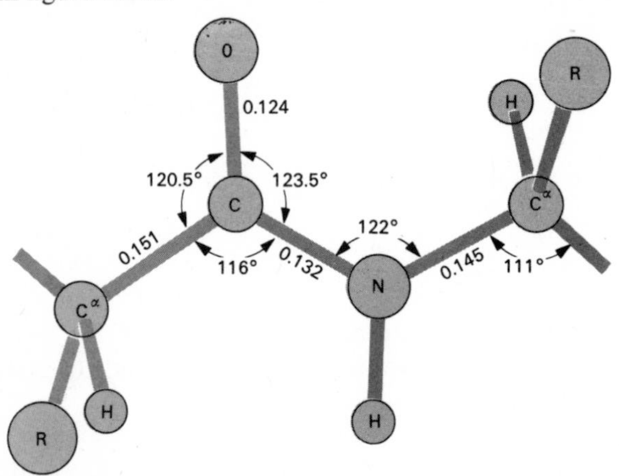

Figure 18.4b
The average dimensions of the peptide group (bond lengths in nm).
From Advances in protein chemistry, **22**, *Academic Press (1967) 249*

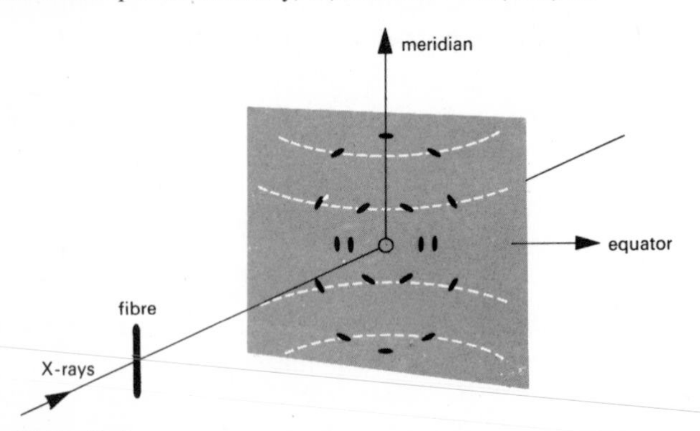

Figure 18.4c
Diffraction by a fibre.
After Wilson, H. R. (1966) Diffraction of X-rays by proteins, nucleic acids and viruses, *Edward Arnold*

X-ray fibre photographs

In the study of long rod-like molecules a special arrangement is often used when taking diffraction photographs. A specimen of material is placed in the

X-ray beam so that the molecules are orientated perpendicular to the X-ray beam. The arrangement is illustrated in figure 18.4c.

Such an arrangement produces what are called fibre photographs.

If a diffraction spot occurs on the vertical axis of the diffraction pattern, it is known as a meridional reflection; if it occurs on the horizontal axis it is known as an equatorial reflection.

If the polypeptide chain of a protein is arranged so that the molecule is rod-like, then taking X-ray photographs of the fibre type may give valuable information about the structural details of the polypeptide chain.

The type of diffraction photograph which might result must now be considered. If the peptide groups occur at regular intervals up the rod-like molecule, they will cause diffraction in a simple way analogous to optical diffraction caused by a two-dimensional grating. Use the Nuffield diffraction grid with the rectangular pattern of dots to view an ordinary pearl light bulb from a distance of about three metres (not a point source; the reason for this will be discussed later). What type of pattern is produced? How does this compare with the diffraction pattern obtained from porcupine quill, figure 18.4d?

For the Nuffield grid the horizontal layer lines of the diffraction pattern can be considered as due to a regular repeat (the dots) on the vertical axis of the grid. What does the occurrence of layer lines in the diffraction pattern of porcupine quill imply about its molecular structure? Look again at the DNA fibre photograph in Topic 8. Are there layer lines of diffraction spots present?

(If the polypeptide chain of a protein is rolled up in some complex way, then the protein molecule will be globular rather than rod-like and quite different X-ray patterns will result. See for example the X-ray crystal diffraction photograph of lysozyme illustrated in figure 18.5c.)

The structure of polypeptides

The information available in the diffraction pattern of naturally occurring proteins with rod-like molecules is insufficient to determine their structures by direct calculation. There is, for example, the possibility that the rods will not be regularly arranged (see figure 18.4e) and this will lead to some diffuseness in the diffraction pattern obtained (for this reason the Nuffield grid was viewed with a diffuse light source so as to mimic more closely the results obtained with natural materials).

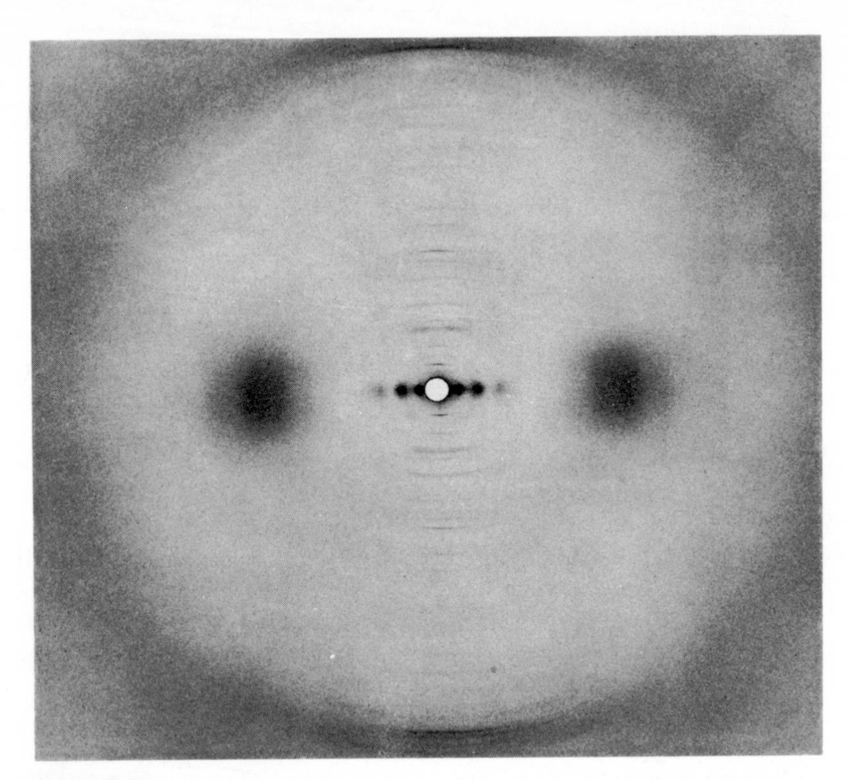

Figure 18.4d
X-ray diffraction photograph of quill tip of the South American porcupine (Coendu paraguayensis). The photograph is much enlarged.
After Cohen, C. and Holmes, K. C. (1963) J. Mol. Biol. **424**, *6*, *Academic Press*

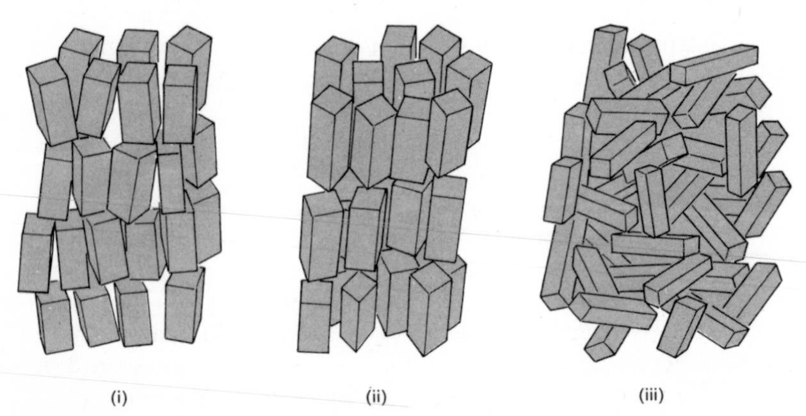

Figure 18.4e
Different types of disorder **i** mosaic **ii** texture **iii** random, for rod-like molecules.

Attempts have therefore been made to simplify the problem by studying synthetic polypeptides made from one amino acid only. Valuable information was obtained from the diffraction pattern of polybenzyl glutamate (figure 18.4f).

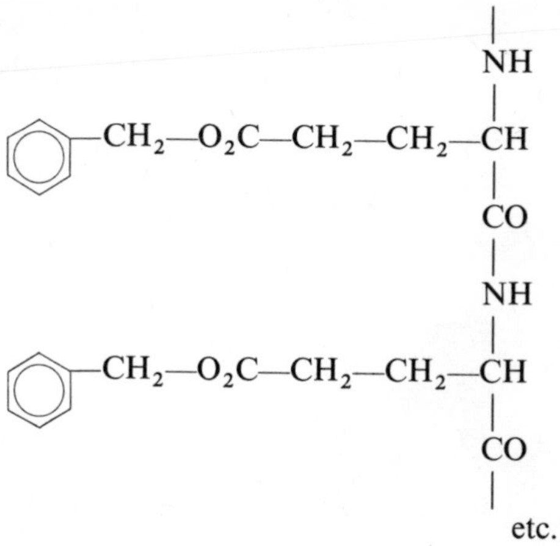

$$
\begin{array}{l}
\qquad\qquad\qquad\qquad\qquad\qquad\quad | \\
\qquad\qquad\qquad\qquad\qquad\qquad\quad NH \\
\qquad\qquad\qquad\qquad\qquad\qquad\quad | \\
C_6H_5{-}CH_2{-}O_2C{-}CH_2{-}CH_2{-}CH \\
\qquad\qquad\qquad\qquad\qquad\qquad\quad | \\
\qquad\qquad\qquad\qquad\qquad\qquad\quad CO \\
\qquad\qquad\qquad\qquad\qquad\qquad\quad | \\
\qquad\qquad\qquad\qquad\qquad\qquad\quad NH \quad \text{polybenzyl glutamate} \\
\qquad\qquad\qquad\qquad\qquad\qquad\quad | \\
C_6H_5{-}CH_2{-}O_2C{-}CH_2{-}CH_2{-}CH \\
\qquad\qquad\qquad\qquad\qquad\qquad\quad | \\
\qquad\qquad\qquad\qquad\qquad\qquad\quad CO \\
\qquad\qquad\qquad\qquad\qquad\qquad\quad | \\
\qquad\qquad\qquad\qquad\qquad\qquad\quad \text{etc.}
\end{array}
$$

The meridional reflection (which lies just inside the diffuse halo) from polybenzyl glutamate can be shown by a simple calculation to correspond to a periodic repeat every 0.54 nm along the fibre axis of the molecule.

Any proposed structure for this polypeptide will have to account for this value.

Another meridional reflection, corresponding to a 0.15 nm repeat along the fibre axis, was first observed by M. F. Perutz.

The problem now is to build a scale model of the polypeptide chain which will have regular repeats every 0.54 nm and also every 0.15 nm. After a plausible scale model has been built, it can be analysed and its diffraction pattern calculated. Comparison with actual diffraction patterns for reflection positions and intensities will help to decide if it is a successful model.

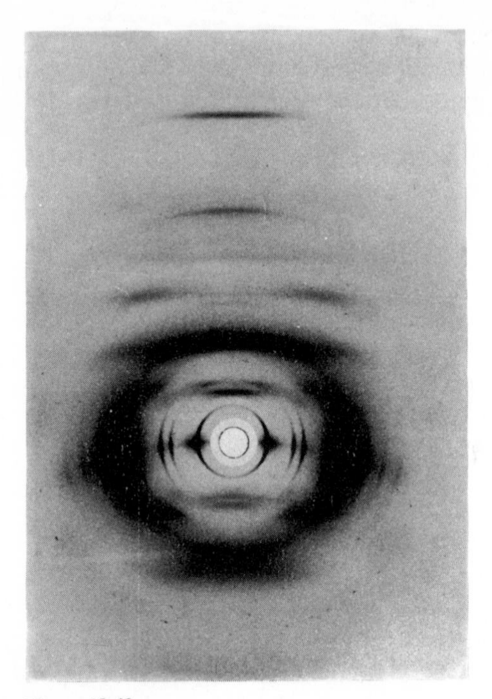

Figure 18.4f
X-ray diffraction photograph of polybenzyl glutamate fibres. The outermost reflection is the 0.15 nm reflection and the large diffuse reflection on the meridian is in the 0.52 nm region.
Photo, Courtaulds Ltd

The α-helix

By giving due weight to all the experimental information the most successful model to be built of the polypeptide chain in rod-like protein molecules was proposed by Pauling and Corey.

They suggested a helical structure of the 18 amino acid residues in 5 turns of the helix, giving $\frac{18}{5} = 3.6$ amino acids per turn. With 0.54 nm per turn of the helix the distance between each amino acid step in the helix is $\frac{0.54}{3.6}\ \text{nm} = 0.15\ \text{nm}$ as required by the X-ray diffraction evidence.

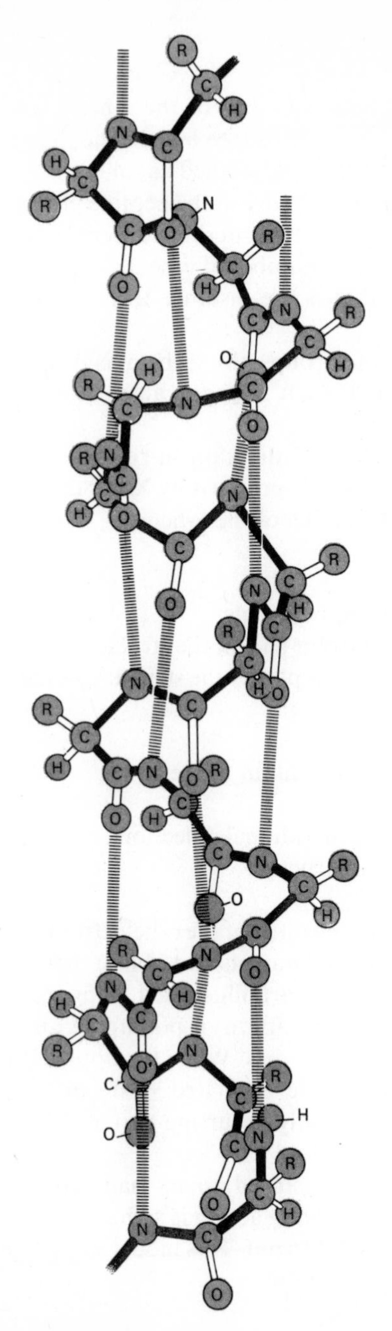

Figure 18.4g
A two-dimensional representation of the α-helix.

Experiment 18.4b

The α-helix

The model proposed by Pauling and Corey is known as the α-helix, and it is part of the structure proposed for fibrous proteins (see later). The illustration (figure 18.4g) is approximately to scale and can be studied as follows.

1 Trace the hydrogen bonding sequences between the peptide groups, and mark (by colouring) the O—C—N atoms of the interlinked peptide groups. Measure the hydrogen bond scale length in a number of cases.

2 Measure the pitch of the helix, that is the distance from wave-top to wave-top.

3 Measure the spacing of the peptide groups along the axis of the helix. To do this draw horizontal lines through each nitrogen atom and measure their vertical spacing.

4 Are the dimensions as measured on the diagram in roughly the same ratio as the experimental nanometre spacings determined by X-ray diffraction?

5 Attempt to build a ball-and-spoke model of the α-helix.

Background reading

The structure of naturally occurring materials

The establishment of the α-helix as an important part of the structural organization of synthetic polypeptides did not solve completely the structure of naturally occurring fibrous proteins.

Look at the diffraction pattern of porcupine quill in figure 18.4h.

A significant detail in the pattern is a strong meridional reflection corresponding to a repeat every 0.516 nm in the molecular structure.

The problem is to explain the shortening of the pitch of the α-helix from 0.54 nm to 0.516 nm. The technique adopted is to build models and obtain their optical diffraction patterns. The results of such work are illustrated in figure 18.4h, where you can compare the X-ray diffraction pattern of porcupine quill with optical diffraction patterns based on various models. While the optical diffraction patterns were recorded, the models were oscillated $\pm 10^{\circ}$ in order to simulate the element of disorder in the naturally occurring material.

The optical pattern (ii) is based on a three-strand rope, made from three α-helixes which coil round each other. This coiling action is sufficient to reduce the pitch within the α-helix from 0.54 nm to 0.516 nm. This model was proposed by Crick in 1953 and is illustrated in figure 18.4h.

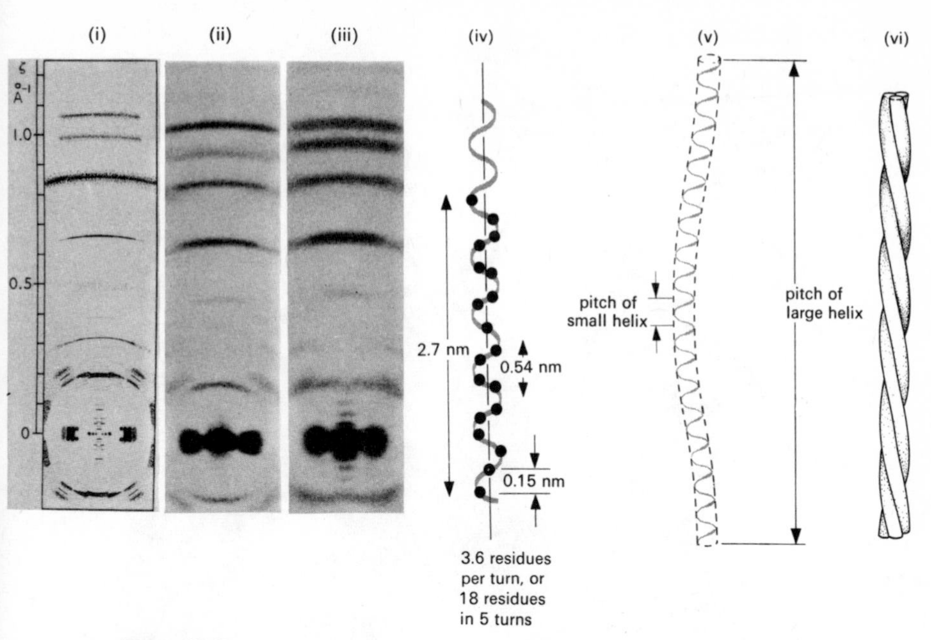

Figure 18.4h
i X-ray diffraction pattern of porcupine quill.
ii Optical diffraction pattern from three-strand rope.
iii Optical diffraction pattern from segmented three-strand rope.
iv An α-helix.
v A compound helix.
vi A three-strand rope or coiled-coil.
Photos, Dr R. D. B. Fraser, CSIRO, Australia

Further improvements have been proposed to allow for the possibility of segments of rope having different amino acid composition, producing portions of rope with straight runs of α-helixes rather than continuously coiling α-helixes. The optical pattern is illustrated at (iii).

Nevertheless the structure of porcupine quill and other α fibrous proteins such as wool are still very imperfectly understood.

An interesting example of a globular protein is myoglobin, whose function is to act as a temporary storehouse in cells for oxygen brought to them by the haemoglobin in blood. Myoglobin was the first protein to have its three-dimensional structure fully appreciated and much of its polypeptide chain has the α-helix conformation (see figure 18.4j).

Figure 18.4j
The structure of the myoglobin molecule shows the several straight sections of α-helix linked by short lengths of non-helical chain. The haem group, with an iron atom at its centre, is held in place between the chains.
After Dickerson, R. E. (1964) The proteins, **2**, *Ed. Neurath, H. Academic Press*

It should be mentioned that there are other (β) fibrous proteins, such as swans' feather and silk which have rather different diffraction patterns; for silks a pleated sheet structure has been proposed by Pauling and Corey. The peptide chains are linked through the sheet by hydrogen bonds and the general structure was illustrated in Topic 10. Look carefully at the space-filling diagrams in figure 18.4k and note how the polypeptide chains are differently packed in two varieties of silk.

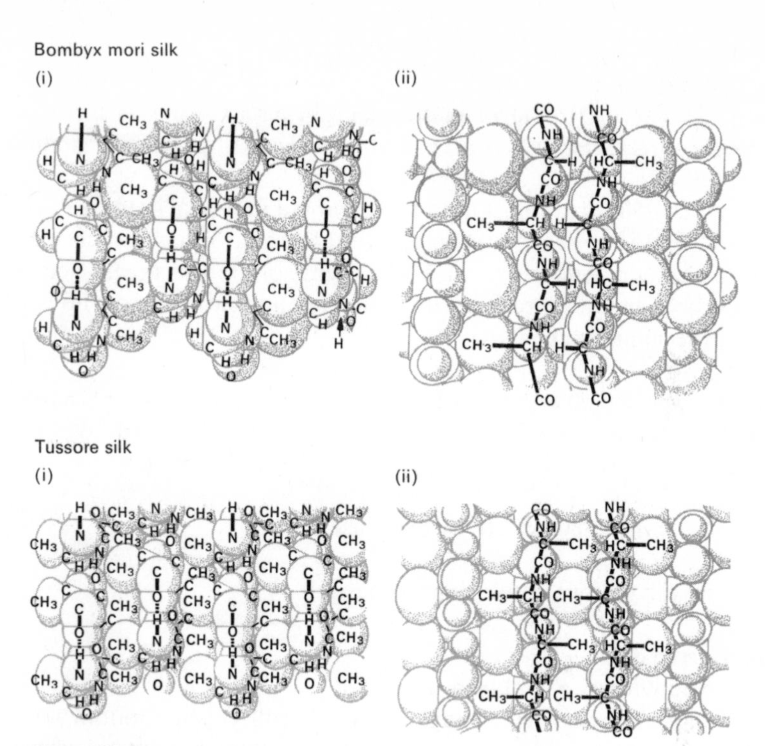

Figure 18.4k
Space-filling drawings of the structure of Bombyx mori and Tussore silk.
i viewed along the fibre axis.
ii viewed parallel to the plane of the pleated sheets.
(*top*) *After Marsh, R. E., Corey, R. B., and Pauling, L.* (*1955*) Biochimica & Biophys. Acta **16**,
(*bottom*) *After Marsh, R. E., Corey, R. B., and Pauling, L.* (*1955*) Acta Cryst. **8**, *713*

18.5 **Enzymes**

Enzymes and their functioning are of fundamental biochemical significance, yet their study is liable to fragmentation because the skills required are not to be found within any of the traditional scientific divisions; physicists, chemists, and biologists have all made important contributions.

Enzymes are proteins whose function in a living organism is to help bring about necessary biochemical reactions. Any compound whose reaction occurs through the intervention of an enzyme is known as a *substrate* of that enzyme.

The following reactions are designed to illustrate qualitatively the properties of enzymes. When carrying out the reactions you should bear in mind the properties of conventional catalysts.

Experiment 18.5a

The specificity of urease

Urease is an enzyme found in plants; jack beans or water melon seeds are convenient sources. It converts urea to ammonia by a hydrolysis reaction.

To 5 cm^3 of a 0.25M urea solution add five drops of universal indicator, followed by drops of 0.01M hydrochloric acid until the indicator has just changed to a distinct red colour. Add 1 cm^3 of a 1 per cent solution of urease active meal (which has been similarly treated with indicator and acid) and note how quickly the pH of the solution changes.

Repeat the experiment with 0.25M solutions of compounds which have structural similarities to urea and might therefore be hydrolysed by urease to ammonia.

How specific is the enzyme activity of urease?

What causes the pH of the solution to change?

Experiment 18.5b

The activity of α-amylase

Saliva contains an enzyme, α-amylase, which will act at random along the polysaccharide chain of starch breaking it down eventually to the sugar maltose. If you chew a piece of bread and retain it in your mouth for five minutes, you should notice a change in flavour. The following experiments are an indication of some of the factors which influence the activity of α-amylase.

To obtain a suitable sample of saliva first rinse your mouth out, then wash it round with 20 cm^3 of warm water for a minute and collect the washings in a beaker. The washings will serve as 'dilute saliva' for the following experiments.

1 To 10 cm^3 of a fresh 1 per cent starch solution add 2 cm^3 of dilute saliva. Every minute for six minutes remove two drops of solution and add them to 1 cm^3 of 0.001M iodine solution. Note the deterioration of the starch-iodine blue coloration as the starch is degraded.

2 To test the influence of pH repeat experiment (1) but add either 1 cm^3 of 0.1M hydrochloric acid or 1 cm^3 of 0.1M sodium hydroxide before adding dilute saliva.

3 To test the effect of temperature stand samples of starch solution in water baths at 40 °C and 95 °C. When the starch solutions have heated up add dilute saliva and test as in experiment (1).

With care experiments (1) and (2) can be run simultaneously, especially if samples are only tested after one, two, four, and six minutes.

What are the conditions of pH and temperature at which saliva functions in your mouth? Do your experimental results support these as being good conditions for saliva activity?

What went wrong at 95 °C? Is there an additional experiment you can perform to decide which failed, the starch or the saliva?

How could you test starch–saliva mixtures for the presence of sugars?

Experiment 18.5c

The digestion of milk

Milk is an important foodstuff and to get maximum food value it is attacked in the digestive system by a variety of enzymes of which two are investigated here.

Rennin is an enzyme found in the lining of the stomach. It clots milk thereby producing a solid surface on which other enzymes can act to digest the protein in milk. It is used in the kitchen to make junket.

1 To 10 cm^3 of milk in a water bath at 40 °C add 1 cm^3 of rennin solution and note the consistency of the mixture after five minutes.

2 Carry out the same experiment but first remove calcium ions from the milk by adding 3 cm^3 of a 0.1M oxalate solution and filtering to remove the calcium oxalate precipitate.

3 To study the effect of heat on the enzyme repeat the first experiment but, before you add the 1 cm^3 of rennin solution, heat it until it just boils.

Lipase is an enzyme found in the pancreas. It breaks down fats in alkaline solution producing the free fatty acids.

1 Briefly boil 10 cm^3 of milk to destroy any enzymes present. Cool and add three drops of phenolphthalein solution followed by drops of dilute sodium carbonate solution until there is a distinct pink colour.

2 Divide the alkaline milk into two portions and place in a water bath at 40 °C. As a source of lipase use a 1 per cent solution of pancreatin: to the portions of milk add (*a*) a 1 cm^3 portion; (*b*) a boiled 1 cm^3 portion.
What happens? What explanation can you offer?

Substrate	Experimental conditions (room temperature unless stated)	Enzyme
18.5a		
5 cm^3 urea	pH 2	1 cm^3 urease
5 cm^3 thiourea etc.	pH 2	1 cm^3 urease
18.5b		
10 cm^3 starch	untreated	2 cm^3 dilute saliva
10 cm^3 starch	1 cm^3 0.1M HCl	2 cm^3 dilute saliva
10 cm^3 starch	1 cm^3 0.1M NaOH	2 cm^3 dilute saliva
10 cm^3 starch	40 °C	2 cm^3 dilute saliva
10 cm^3 starch	95 °C	2 cm^3 dilute saliva
18.5c		
10 cm^3 milk	40 °C, untreated	1 cm^3 rennin
10 cm^3 milk	40 °C, calcium ions removed by addition of oxalate	1 cm^3 rennin
10 cm^3 milk	40 °C, enzyme boiled before use	1 cm^3 rennin
5 cm^3 milk	40 °C, milk boiled and made alkaline	1 cm^3 lipase
5 cm^3 milk	40 °C, milk boiled and made alkaline, enzyme boiled before use	1 cm^3 lipase

Table 18.5a
Summary of enzyme experiments

As a result of your experimental work you should have established a number of ideas about the properties of enzymes.

Unlike a conventional catalyst such as platinum which catalyses a variety of reactions, enzymes are highly specialized and often for a particular enzyme there is only one reaction of one substrate which it can catalyse. That is, enzymes are *specific* catalysts. See chapter 9 in *The chemist in action* for further reading on catalysts.

As an example of the specific activity of enzymes, fumarase catalyses only the addition of water to fumaric acid, and, furthermore, of the two possible optically active products only one isomer is produced:

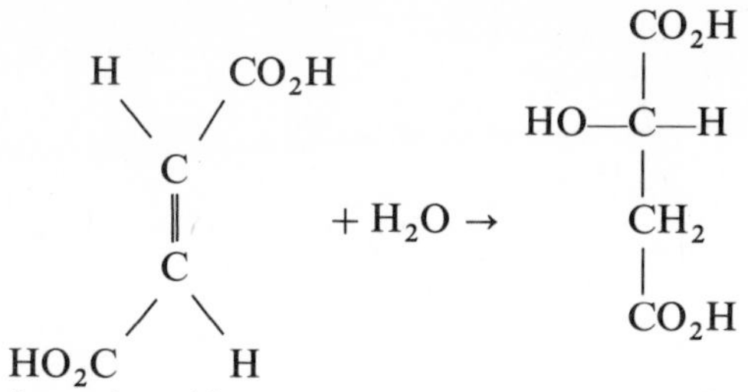

$$\begin{array}{ccc} \mathrm{H} & & \mathrm{CO_2H} \\ & \diagdown\ \diagup & \\ & \mathrm{C} & \\ & \| & \\ & \mathrm{C} & \\ & \diagup\ \diagdown & \\ \mathrm{HO_2C} & & \mathrm{H} \end{array} + H_2O \rightarrow \begin{array}{c} \mathrm{CO_2H} \\ | \\ \mathrm{HO{-}C{-}H} \\ | \\ \mathrm{CH_2} \\ | \\ \mathrm{CO_2H} \end{array}$$

fumaric acid — L-malic acid

Fumarase is also highly efficient, with a conversion rate at 25 °C of 10^3 molecules of substrate every second per molecule of enzyme! In fact the reaction is so fast, it is controlled by the rate at which the molecules meet by diffusion, i.e. the enzyme causes the substrate to react every time they collide.

The influence of temperature on catalyst activity is considerable. If the temperature is too high, or too low, enzymes cease to function. This leads to the concept of an *optimum temperature* as being the temperature at which, for a given set of conditions, the enzyme will catalyse the greatest amount of chemical change.

Thus a digestive enzyme from a sea squirt was found to have an optimum temperature of 50 °C for a two hour reaction period. Yet being in a sea animal the enzyme will be required to function at a temperature of 15 °C. This does not seem a very efficient digestive situation for the sea squirt until it is realized that it takes up to 60 hours to digest its food. When the optimum temperature over a 60 hour period was investigated it was found to be 20 °C. This illustrates the care needed for the proper investigation of enzyme properties.

What reason can you suggest for enzymes being deactivated by high temperatures?

Enzymes also function best at a particular pH, the *optimum pH*, which is not usually very different from neutrality, being mostly in the range pH 5–7. By the use of buffer solutions the pH dependence of α-amylase from saliva can be more closely studied (figure 18.5a).

How might the structure of an enzyme be influenced by pH? Remember that enzymes are proteins, so amino acids such as lysine and glutamic acid (see table 18.2b) may be present.

An initial understanding of the mechanism of enzyme activity was obtained by kinetic studies.

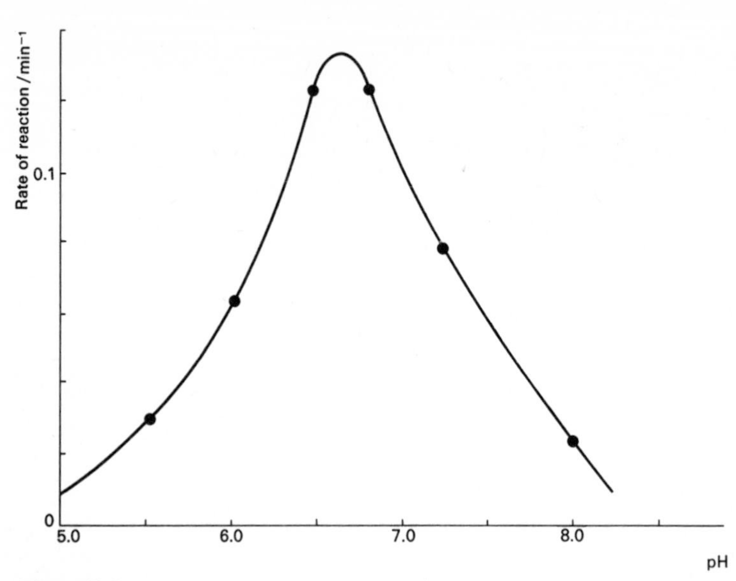

Figure 18.5a
Hydrolysis of starch by saliva (student result).

The results suggest that enzymes act catalytically on their substrates and that their reaction, for example for addition of water to the substrate, might be represented:

enzyme + substrate → enzyme-substrate complex
complex + water → enzyme + hydrated substrate.

The last point to be considered is the exact nature of the substrate-enzyme complex and how it contributes to the breaking and making of bonds in the substrate. In the case of lysozyme and its substrate Professor D. C. Phillips and his co-workers at Oxford have proposed an answer.

Background reading

1 The structure and activity of lysozyme

Lysozyme was first noted in human nasal mucus by Fleming but a more convenient source is hen egg-white. Its specific function is to break down a polysaccharide found in the cell walls of certain bacteria, thereby killing the bacteria. Chemical studies have established that its molecule consists of 129 amino acids in a single chain, crosslinked in four places by disulphide bridges (figure 18.5b). The overall shape is globular and X-ray diffraction studies have revealed the detailed conformation of the amino acid chain.

Figure 18.5b
The amino acid sequence of lysozyme.

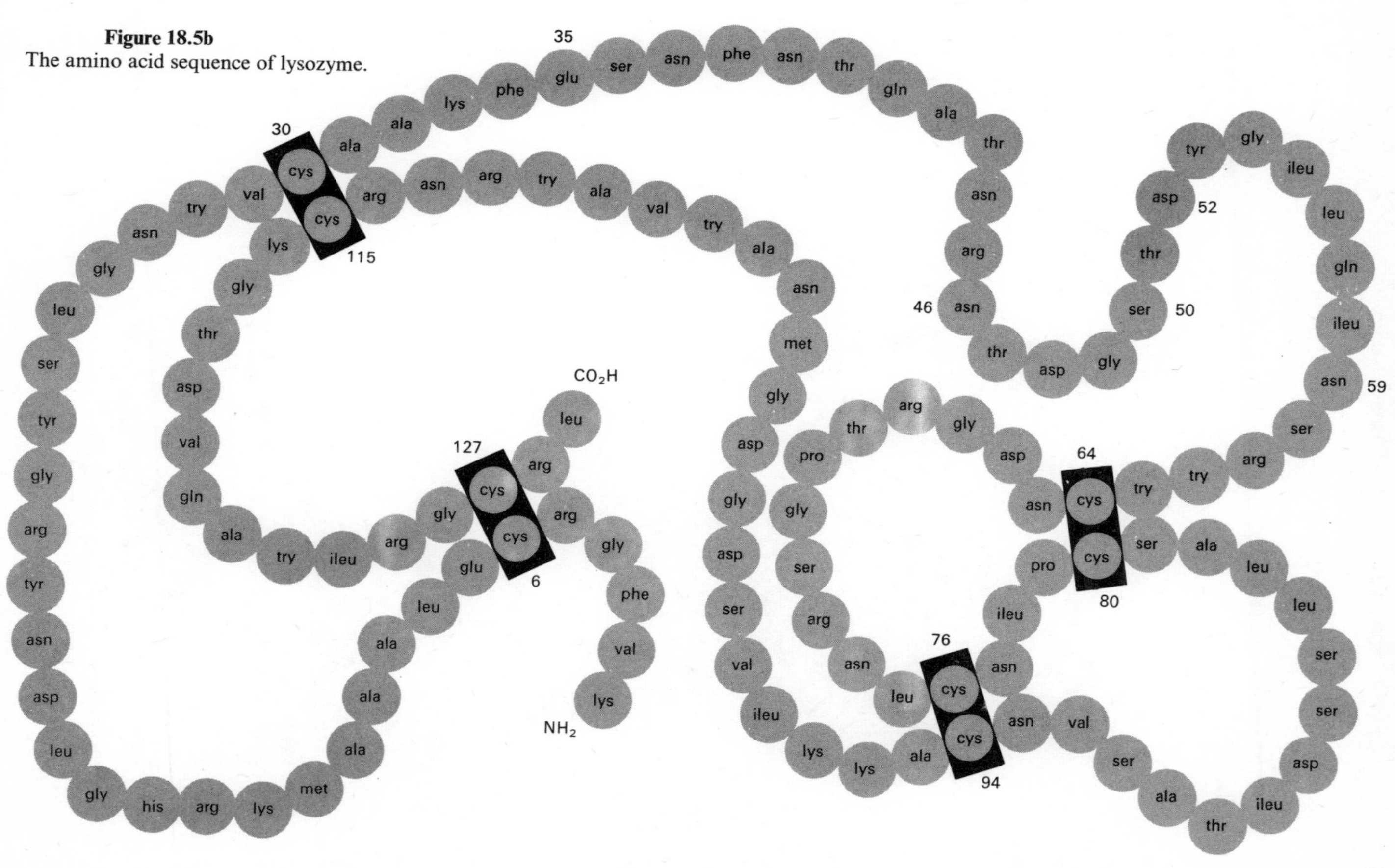

A typical diffraction photograph is illustrated in figure 18.5c and below that a small portion of the electron density map (figure 18.5d). At b you are looking down through a short run of α-helix, and at c the heavy peak is due to a disulphide bridge.

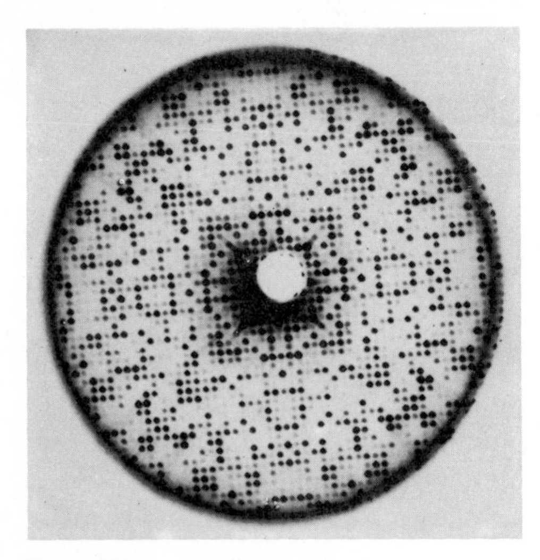

Figure 18.5c
X-ray diffraction photograph of lysozyme.

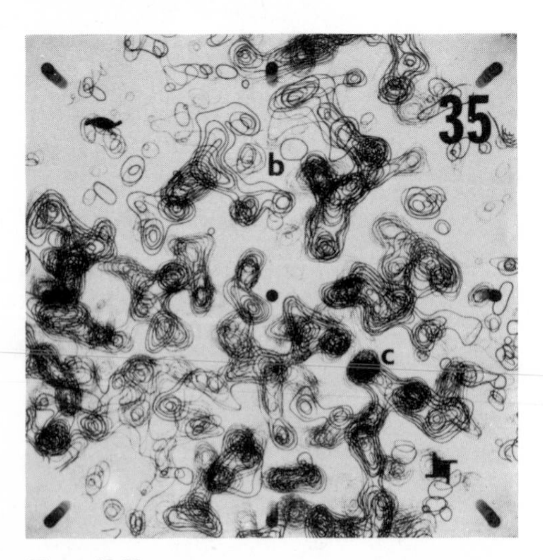

Figure 18.5d
Part of the electron density map of lysozyme. This should be compared with maps of similar molecules.

As well as the diffraction pattern of lysozyme it is also necessary to take photographs of crystals to which a heavy atom substituent has been added, without disturbing the protein structure. (This is known as the isomorphous replacement method.) The calculations require a high speed computer and even then it is usual to develop the structure in stages. At first only the central part of the diffraction pattern, corresponding to reflections down to 0.6 nm, is used in calculations. The structure deduced is illustrated as the 0.6 nm model (figure 18.5e). The most impressive feature is the deep cleft in the centre of the molecule.

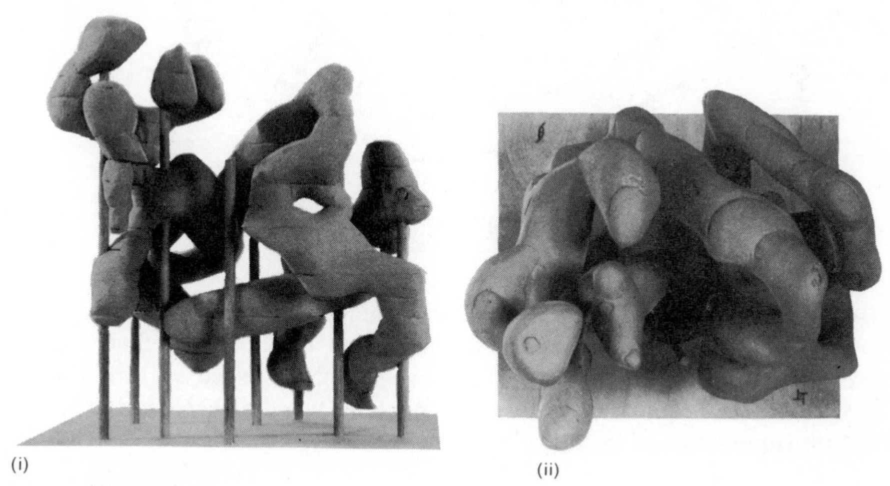

(i) (ii)

Figure 18.5e
0.6 nm resolution model of lysozyme.
i front view **ii** view from above, looking down into the active site.

As well as lysozyme itself the diffraction patterns of lysozyme-substrate complexes have been studied and it has been found that polysaccharides are bound in the deep cleft in the molecule.

Lysozyme is known to split the bond between two sugar residues. By considering reflections down to 0.2 nm it has been possible to make detailed proposals about the action of lysozyme on this substrate. The detailed conformation of lysozyme is illustrated in figure 18.5f and its action on the cell wall polysaccharides of bacteria is discussed in chapter 9 of *The chemist in action.*
The lysozyme structure is designed with great precision to bind a particular polysaccharide so that one of its bonds can be broken by being placed alongside those amino acid residues which have the right properties and energy.

In conclusion it is worth noting the increasing rate of discovery of protein structure. Perutz took his first X-ray diffraction pictures of haemoglobin in

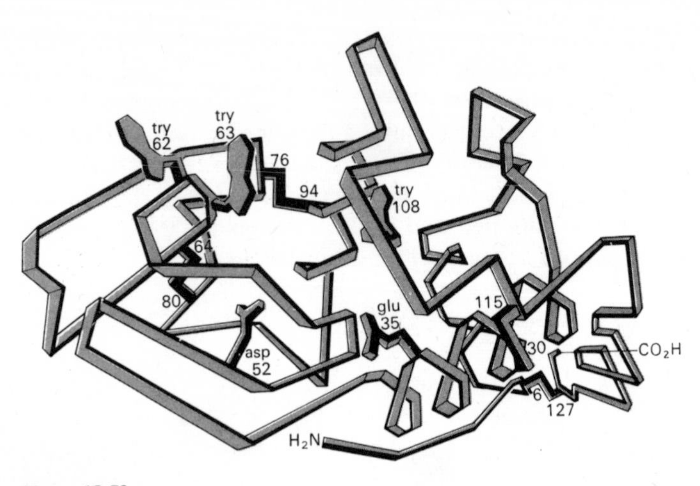

Figure 18.5f
The three-dimensional structure of lysozyme (schematic).
Figure 18.5 b–f, by courtesy of Professor D. C. Phillips, F.R.S.

1937, but the isomorphous replacement technique which is vital for the interpretation of the diffraction patterns was not found until 1953. The first detailed protein structure worked out was for myoglobin, published by Kendrew in 1958. The closely related structure of haemoglobin was discovered in 1959. This was the culmination of 25 years pioneering work for which Perutz and Kendrew were awarded a Nobel prize in 1962.

Since then the pace has accelerated, work on lysozyme started in 1960 and the structure was published in 1965. Now new enzyme structures are published nearly every year.

2 Enzymes for industry

Chemical reactions can be catalysed in a number of ways. Finely divided metals such as platinum, nickel, and palladium are capable of catalysing a wide range of reactions, especially at very high temperatures. Acids and bases are also used as chemical catalysts. Living organisms, however, cannot survive extremes of acidity or alkalinity or high temperatures, so that enzymes – the catalytic proteins of living organisms – must be active under much more moderate conditions than those used in non-living chemical catalysis.

However, despite these limiting factors, enzymes are between one hundred thousand and one hundred million times more efficient than other chemical catalysts. One molecule of a particular enzyme can promote a reaction in as many as ten million molecules of the substrate it attacks in one second. And enzymes have an even more important advantage over non-biological catalysts:

whereas the latter catalyse a variety of reactions, an enzyme catalyses only one kind of reaction – the specific conversion of a particular substance, the substrate, to a particular product.

A single living cell may contain as many as one thousand enzymes, each controlling a specific reaction. The potential usefulness of enzymes in industry is therefore enormous, but, because purified enzymes are very expensive – the majority cost more than £10 per gramme – and they usually have to be destroyed or discarded after they have been used only once, only a few are actually employed by industry on a large scale. Table 18.5b presents a summary of most of the industrial processes which employ added enzymes. Other industries, notably the brewing industry, are of course dependent upon enzymes – fermentation is an enzymic process – but the enzymes in these cases are produced naturally by the material involved; they are not 'added enzymes', and so are not included here.

The added enzymes used in industry are normally of plant or microbial origin. Papain, for example, is readily obtained from papaya fruit. Certain bacterial enzymes may be easily isolated from the medium in which the bacterial culture is grown, since they are secreted into the medium by the bacteria. However, although bacteria and other micro-organisms contain many useful enzymes, only a few are secreted into the culture medium, so as to become available: normally the cell walls of these organisms would have to be disrupted to liberate enzymes. Large-scale methods for accomplishing this are inadequate at present and the difficulty has yet to be overcome.

Enzymes are natural products and are acceptable as food additives, a fact of vital importance for the food industry where they find extensive use. The popular enzymes are, of course, the cheap ones, easily isolated from extremely cheap sources. They are not needed in pure form and some loss of enzymic activity during preparation and storage can be accepted. The instability of enzyme activity can even be useful. For instance, the enzymes used in bread-making and meat-tenderizing are destroyed by cooking, so that their activity is automatically ended at the appropriate stage in the preparation of the foods.

A few enzymes have a multitude of uses. Papain, for example, will break down the proteins in beer which would otherwise form protein-tannin precipitates (hazes) when the beer is chilled. This is called 'chillproofing'. Another use, in tenderizing poor cuts of meat, involves the splitting of the protein molecules of the muscle. Amylases are widely used to remove starch, for instance by breaking it down to the sugars needed for fermentation by the yeast in bread-making, and for removing starch-based thickening agents used in connection with the printing of cotton and rayon goods (desizing process). Large-scale production of glucose can be achieved from starch (corn, wheat, or potato in origin) using an amylase of fungal origin together with another enzyme called amyloglucosidase to allow the breakdown to go to completion, giving glucose as the product.

An interesting example of an enzyme used as a food additive is invertase, which converts sucrose (table sugar) to glucose and fructose (fruit sugar). The enzyme is mixed with sucrose in the manufacture of soft-centred chocolates. The mixture is coated with chocolate while it is still solid, and only later does liquefaction occur as the enzyme splits the sucrose to the more soluble products, which dissolve in the small amount of water present. The enzyme is not destroyed in this case, but does no harm when it is eaten. In any case, along with almost all added enzymes, it will of course be inactivated upon contact with the hydrochloric acid of the stomach.

The ease with which enzymes can be destroyed has its commercial value, but many enzymatic processes could be run continuously, as is often the case with other catalysts, if only the enzyme was not destroyed during the manufacturing process. The isolation of uncontaminated products of a reaction often requires the removal of enzyme through precipitation by heat, acidification, or organic solvents. Unfortunately all of these methods destroy enzyme activity. Suitable techniques might be available to recover an active enzyme from the final products, but this would be an expensive process. Here lies the need for an enzyme which can be recovered, 'alive', at the end of the process, and which can be used again.

Enzymes are now available in a new form, in which the enzyme is bound by incorporating it into an insoluble synthetic polymer or reacting it with specially treated cellulose (table 18.5c). Although slightly less active in this form, in most cases the resulting bound enzyme is more stable towards heat and extremes of acidity or alkalinity and is less susceptible to oxidation (a common cause of enzyme inactivation) than the corresponding natural enzyme. However, the most important feature of bound enzymes, as far as industry is concerned, is that they can easily be recovered from the reaction mixtures, either by filtration, or centrifugation, and used again – often repeatedly. The effective cost of using such enzymes in industry is thus greatly reduced.

Bound enzymes, then, are a new type of product which appear to have a great industrial potential. At the moment there are three main ways in which enzymes are used: first, the natural enzymes of the material may be used, for instance, fruit juices are often clarified by protein-splitting enzymes already present in the juices; secondly, enzymes may be added to act on a solid material, as in the tenderizing of meat by papain; and thirdly, enzymes may be added to liquids. It is in this last method that bound enzymes can be used.

Name	Industry	Uses
Amylases		
(split starch)	brewing	Liquefaction of added cereals
	textile	Removal of starch (sizes) added during processing
	baking (bread)	Starch → glucose. Yeast action on glucose produces carbon dioxide. Get better colour and gassing powers during baking – also retard staling
	paper	Used to degrade starch to lower-viscosity product needed for sizing and coating processes
Proteases		
(split proteins)		
Papain	brewing	To chillproof beer
	meat	To tenderize meat: the action of this fairly heat-resistant enzyme can continue while the meat is warming. The enzyme is of course inactivated as cooking proceeds
	pharmaceutical	Used in tooth powders to remove protein deposited on teeth
Ficin	photographic	To dissolve gelatin of scrap film allowing recovery of silver present
Pepsin	cereal	In production of instant cereals
	pharmaceutical	As digestive aid – supplement to normal action of pepsin in stomach
Trypsin	food	To predigest baby foods
Rennin	cheese	To precipitate the protein (casein) from milk
Bacterial		
proteases	laundry	Enzyme-containing washing powders
	leather	Loosening of hair without damage to hide or hair. Leather treatment (bating)
	textile	Recovery of wool from sheepskin pieces
	food	Preparation of hydrolysed proteins, e.g. for animal feedstuffs.

Table 18.5b
Summary of some common industrial use of enzymes

Bound Enzyme	Support material	Enzyme: support (approximate ratio)
Chymotrypsin	synthetic polymer	1:1
Ficin	cellulose derivative	1:10
Papain	cellulose derivative	1:10
Trypsin	cellulose derivative	1:10
Trypsin	synthetic polymer	1:4

Table 18.5c
Some commercially available bound enzymes.
New Scientist, *April* 11*th*, 1968.

18.6 An examination of some plastics

Plastics are materials to be used, not chemical curiosities. In many of their applications they are superior to all other possible materials. What material other than polythene is simultaneously tough, flexible, free from rot and rust, light-weight, and cheap? In order to appreciate the versatility of plastics consider some of the many ordinary household objects that are made of plastics, find out what they used to be made of (if they were made at all), and decide which is the better material in terms of convenience and cost. Some information which may be useful is given in table 18.6.

Material	Cost in 1968 /p dm^{-3}	Weight /g dm^{-3}
Nylon	66	1100
Aluminium	60	2600
Steel	35	7800
Polythene	16	930
P.V.C.	14	1300
Cardboard	4	720
Softwood	4	560

Table 18.6
Comparative cost and weight for some common materials

Although a plastic may be the best material for a particular application, the manufacturer is not faced with a simple choice such as enamelled iron or plastic for a bucket, but a choice between iron or polythene or polypropylene or polyvinylchloride or many other plastics, each with distinctive characteristics which must be evaluated before a proper choice is made. If the wrong choice is made, the bucket may collapse when full of hot water. To overcome such difficulties, there are available commercially many grades of each basic plastic and all have been rigorously tested for chemical and mechanical stability.

Thus 'accelerated wear' tests are used to gauge the effect of long-term use, for example exposure to ozone is used to estimate more quickly the effect of atmospheric oxygen, and ultraviolet irradiation is used to estimate the effect of sunlight.

The cost of a plastic is an important consideration also. Since objects, for example a door-knob, often have to be manufactured to a definite size irrespective of the material used the cost by volume must be borne in mind. Some values are quoted in table 18.6.

You should read the chapter on the polymer chemist in *The chemist in action* (chapter 4) where the development is described of a plastic suitable for micro-groove records.

Experiment 18.6

Choosing a plastic

Consider the use of a plastic to make a familiar object such as a carrier bag (commonly made of paper) or a chemical reagent bottle (commonly made of glass). What characteristics must the plastic have to be suitable? What disadvantages of the commonly used materials have to be overcome? Make notes on the chemical properties, physical properties, and manufacturing process that will be required.

When you have written a specification that the plastic must meet to be successful, the next step is to test experimentally a variety of plastics. The first three tests are an investigation of the stability of plastics to a small range of chemicals. Your objective should be a comparative rating of the plastics (if you wish aluminium foil could be similarly tested). Test each plastic the same way in turn. Plastics in the form of small granules are suitable for these tests.

The remaining tests are concerned with the mechanical properties of plastics. Because these properties can be varied for a given plastic by the use of additives, only provisional opinions should be formed about the relative merits of the different plastics. For example, a rigid material like unmodified polyvinylchloride can be converted by suitable additives to a very successful flexible sheet. Only qualitative tests are proposed; if you have already carried out some of these tests quantitatively as part of a physics course you should look up the results obtained. Plastic film is best for these tests.

1 Heat one plastic granule on a piece of broken crucible in a *low* Bunsen burner flame. Try to notice the ease of melting and the nature of the flame when it burns.
Caution. Nylon should be heated only in a *fume cupboard.*
PTFE *must not be heated* because very poisonous fumes are evolved. A plastic which can be melted without decomposition is called a *thermoplastic.* Those which char without melting are known as *thermosetting.* They have usually been irreversibly solidified in the final stage of manufacture.

2 Boil a granule briefly with potassium permanganate solution. Then wash the permanganate out of the test-tube and look for signs of reaction on the granule.

3 Test the granules for resistance to solvents by allowing them to soak in (*a*) tetrachloromethane, and (*b*) toluene.

4 Test pieces of plastic film for resistance to abrasion by rubbing with a household scouring powder.

5 Test strips of plastic film for elasticity by attempting to stretch them by hand. Pull the film slowly but steadily, trying to avoid any sudden jerk, and take note of any changes which may take place both in the appearance of the film and in the force required to stretch it, and to break it.
If, after being subject to a stress, a specimen regains its original size and shape it is said to be perfectly *elastic*; if it entirely retains its strained size and shape it is said to be perfectly *plastic.*
It is instructive to examine the effect of heat on specimens while they are being stressed. Tie one end of a strip to a 500 g mass and suspend it from a clamp stand. Now warm the strip *gently* with a Bunsen burner flame. The effect on rubber is worth observing. Observations are best conducted on rubber strip which has been strained to about double its natural length.

6 Carry out a long term corrosion test by burying pieces of plastic film in the ground and leaving them for next year's class to dig up.

7 What fabricating processes can be used with the various plastics? What processes will be suitable for making the article you have under consideration? Using the information you have available on stability to chemicals, mechanical strength, manufacturing processes, and cost of materials, which plastic is most suitable for your application?

18.7 The laboratory preparation of some plastics

A selection of experiments is included and you should only attempt one or two of interest, depending on the availability of time and materials. For the experiments carried out you should identify the monomer and polymer formulae, determine whether addition or condensation polymerization has occurred, classify the product as thermoplastic or thermosetting, and try to identify it with a commercial product.

To understand the technical terms used, section 18.8 should be consulted.

Experiment 18.7

The preparation of plastics

1 The polymerization of styrene

The behaviour of catalysts used to bring about polymerization reactions is such that they might better be described as 'initiators' for they participate in the reaction and are quantitatively used up. In an opposite sense, substances which react to prevent polymerization are known as 'inhibitors'. Read the label on the stock bottle of styrene and note down the inhibitor present.

Styrene is readily polymerized by a catalyst such as lauroyl peroxide.

With the apparatus in a fume cupboard mix 5 cm^3 of styrene and 5 cm^3 of 1,2-dibromoethane in a 150×25 mm test-tube and add roughly 0.1 g of a catalyst. Plug the mouth of the tube with cotton wool and heat in a hot water bath for twenty minutes.

Allow it to cool, dilute with 10 cm^3 of acetone, and decant from any solid. Precipitate the polystyrene by adding an excess of ethanol to the solution, leave to settle, and pour off the liquid. Knead your lump of product with a spatula and leave overnight to dry.

2 The polymerization of acrylamide

Caution. Acrylamide is a skin irritant. You should wear protective gloves for this experiment.

Make a solution of 10 g of acrylamide in 50 cm^3 of water and warm in a 250 cm^3 beaker to *not more than* 85 °C. Pour out into a second 250 cm^3 'throw-away' container (such as a tin can) on an asbestos mat and add about 0.1 g of a persulphate to initiate the polymerization.

Caution is necessary for the reaction is exothermic. Use your data book to estimate the heat of polymerization.

3 The preparation of formaldehyde resins

The development of formaldehyde resins illustrates the difference, at the turn of the century, between an academic chemist to whom an insoluble infusible product was a material that ruined his apparatus, and a more broad-minded chemist such as Baekeland who could see in such a material worthwhile applications.

Caution. These experiments should be carried out in a fume cupboard as formaldehyde fumes are unpleasant.

a Place 20 cm^3 of 40 per cent formaldehyde solution in a 100 cm^3 throw-away container and add about 10 g of urea with stirring until a saturated solution is obtained. Add a few drops of concentrated sulphuric acid, stirring continuously during the addition. When the reaction is complete, wash the residue well with water and dry.

b Place 25 cm^3 of 40 per cent solution ('formalin') in a 100 cm^3 throw-away container and add roughly 10 g of phenol. *Caution. Phenol is caustic and blisters the skin.* Mix thoroughly by stirring, add 2 cm^3 of concentrated sulphuric acid cautiously with stirring. Keep the mixture for several days until condensation is complete.

c If time permits, this longer and more complex experiment can be attempted. It corresponds more accurately to the controlled conditions necessary to get a usable product.

50 g phenol, 100 cm^3 of 40 per cent formaldehyde solution, and 0.6 g sodium hydroxide are added to a round-bottom bolt-head flask and a water-cooled reflux condenser is then fitted. The contents are carefully heated at first and then allowed to reflux for approximately one hour, to give a clear viscous liquid. After partial cooling, a 3 cm^3 portion of lactic acid is added. The flask is placed in a water bath and is fitted with an air leak and condenser for vacuum distillation. Water is removed with the bath at 75 °C, and a pressure of 20–30 mmHg (water pump), and distillation is continued until little if any distillate comes over. The syrup-like resin may now be cast into moulds for the final stage of cross-linking.

Thin-walled test- or specimen-tubes may be used as moulds; if carefully made, satisfactory open-top moulds may be formed from heavy metal foil that has been folded around an object of simple shape. The hot resin syrup is poured carefully into a series of small moulds, care being taken to prevent the entrainment of air bubbles. The filled moulds are then placed in an oven at 70 °C.

(This stage is the 'Bakelite' process which originated the trade-mark.) A moulding should be removed at daily intervals and, when cold, examined. The early preparations will be found to be brittle, but after a few days curing, the resin becomes hard and strong: finally, if over-cured, the resin again becomes brittle. If glass moulds are used, care must be taken when cracking the tubes prior to removing the castings. The translucency of the cast resins depends on the water content of the resin.

4 The 'nylon rope trick'

An impressive demonstration of the formation of a polyamide polymer is easily carried out by allowing the constituent molecules to react at the interface between two immiscible solvents.

This experiment was originated by P. W. Morgan and S. L. Kwolek of E. I. du Pont de Nemours and Co. Inc. and was published in *J. Chem. Educ.*, 1959, **36,** 182 and 530.

Prepare a solution of 1.5 cm^3 of sebacoyl chloride in 50 cm^3 tetrachloromethane in a 250 cm^3 beaker and, separately, a solution of 2.2 g of 1,6-diaminohexane (*caution* – this is caustic) and 4 g of sodium carbonate in 50 cm^3 of water.

Clamp the beaker, and above it clamp the roller system (made from tubing) as shown in figure 18.7. Allow about 2 metres drop from the roller to the receiver.

Now pour the aqueous solution carefully on to the tetrachloromethane solution and, using crucible tongs, pull the interfacial film out, over the rollers, and down towards the receiver. When a long enough rope has formed, the process will go on of its own accord until the reagents are used up.

To obtain a dry specimen of the nylon polymer, wash it thoroughly in 50 per cent aqueous ethanol, then in water until litmus is not turned blue by the washings.

Is your nylon the '66', '6', or '610' variety?

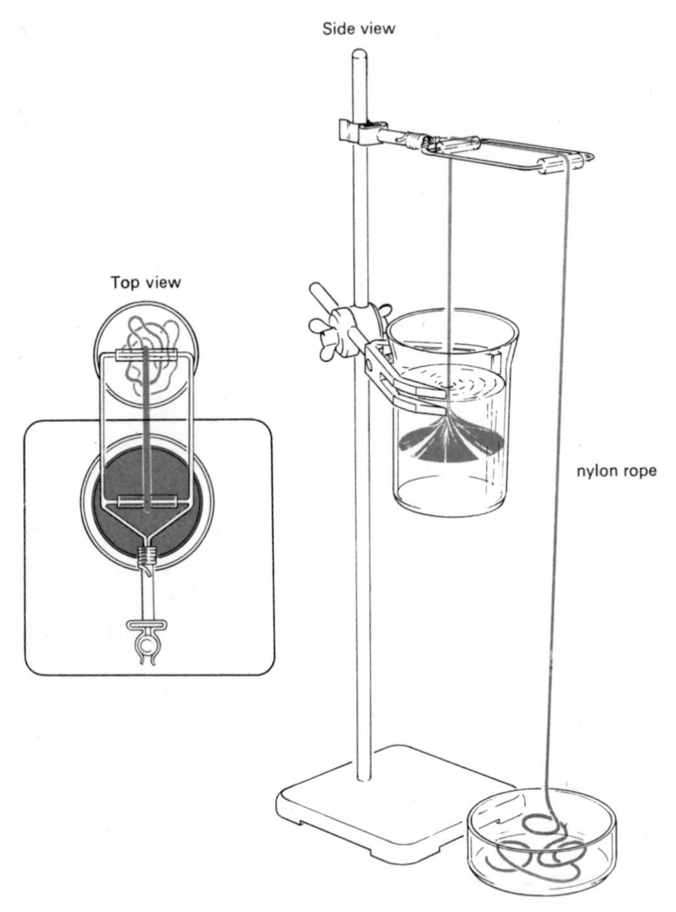

Figure 18.7
The 'nylon rope trick'. (The nylon rope could be pulled out by tongs instead of being allowed to fall.)

5 The production and curing of casein

There have been many attempts to use proteins as raw materials for plastics, and for a time Professor Astbury, a pioneer of the X-ray diffraction of proteins, had a coat made from protein extracted from peanuts.

However at present the only viable product is a casein-formaldehyde polymer mainly used for buttons.

Allow a bottle of milk to stand for a day so that the cream separates. Carefully pour off the top layer and put about 100 cm^3 of the lower layer (separated milk) into a beaker. Warm the separated milk to 50 °C and add 2M acetic acid in drops until no further casein seems to be precipitated. When no further action occurs, remove the lump of casein and press it free from whey with the fingers. Knead it in hot water until it assumes an elastic nature. Wipe it dry and break it in two. Keep one piece as a control and drop the other into 40 per cent formaldehyde solution. Examine it after leaving to stand for a day or two. Dry it and compare the properties of the two pieces. The treated casein should be much harder and bony.

18.8 The invention and discovery of synthetic polymers

Christopher Columbus on his second voyage to America (1493–1496) was perhaps the first European to handle natural rubber; today its structure and properties are still intriguing scientists. Faraday was able to establish the correct empirical formula as C_5H_8 and as early as 1860 rubber was known to be a polymer of isoprene, $CH_2{=}C(CH_3){-}CH{=}CH_2$. However in 1920 the structure was still in dispute, alternative proposals being either a C_{40} ring or a C_{100} long-chain molecule.

German chemists were especially interested in rubber, having produced 2500 tonnes of synthetic rubber during the First World War by the simple expedient of sealing methyl isoprene in iron drums and leaving it for four months to polymerize. It was discovered at that time that sodium would catalyse the polymerization but the technical difficulties were too great to be overcome.

The fundamental structural problems about rubber were solved by the German chemist, H. Staudinger. Pure natural rubber is an unsaturated compound and by addition of hydrogen Staudinger was able to obtain a colourless, inelastic solid, which proved to be a high molecular weight alkane. The hydrogenation conditions were severe; a nickel catalyst was used at 270 °C with a gas pressure of 100 atmospheres. It was in the paper reporting this work that the description 'macromolecule' was first used:

'Rubber is therefore a very high molecular weight hydrocarbon with many olefinic linkages and its chemical behaviour also corresponds fully to this conception. The olefinic linkages can be saturated, partially or completely by the addition of halogens, hydrogen halides or sulphur monochloride by vulcanization, without altering its colloidal properties and therefore without breaking up the 'macromolecules'. In certain reactions the breaking up of the long chain into large or small fragments can occur; for example by the action of ozone or nitrous acid.' (Helvetica Chimica Acta, *No*. 5, 1922, *page* 788.)

However Staudinger's views on the existence of molecules of high molecular weight were not readily accepted, and other chemists supported a 'micelle' theory of conglomerations of molecules held together by van der Waals' forces. Staudinger continued work during the 1920s on X-ray studies and molecular weight methods and finally won acceptance of the view that rubber has molecules with a chain length of the order of $C_{100\,000}$ and a molecular weight of one million.

Once the existence of very long chain carbon compounds had been established and interest was aroused in their properties, the stage was set for the fundamental researches of W. H. Carothers in the Du Pont laboratories at Wilmington, USA. Polymers had returned to the continent where the story started over 400 years before.

W. H. Carothers and the invention of nylon

Wallace Carothers was a brilliant organic chemist, appointed head of a chemistry department before he had even finished the course himself, moving on to Harvard before working for the Du Pont company from 1928 until his premature death in 1937. Carothers' success was based on an accurate understanding of precisely defined objectives '. . . primary object was to synthesize giant molecules of known structure by strictly rational methods' or more informally 'It would seem quite possible to beat Fischer's (molecular weight) record of 4200.'

His early papers clearly state the reasoning behind his work. Thus polymerization is described as '*any chemical combination of a number of similar molecules in which they form a single molecule*. A polymer then is any compound that can be formed by this process or degraded by the reverse process: rubber can be formed by the reaction of isoprene with itself, and cellulose can be hydrolysed to glucose.'

Furthermore he distinguished two types of polymerization reaction:

1 *Addition polymerization* produces a polymer with the same empirical formula as the monomer from which it was made, for example, Neoprene rubber which was first made by Carothers from chloroprene in 1931.

$$\underset{\text{monomer}}{n\mathrm{CH_2{=}C(Cl){-}CH{=}CH_2}} \rightarrow \underset{\text{polymer}}{\left[\mathrm{-CH_2{-}C(Cl){=}CH{-}CH_2-}\right]_n}$$

2 *Condensation polymerization* produces a polymer from monomer molecules by elimination of simple substances such as water or hydrogen chloride, for example, nylon (type 66), first described by Carothers in a 1935 patent.

$$n\mathrm{HO_2C{-}(CH_2)_4{-}CO_2H} + n\mathrm{H_2N{-}(CH_2)_6{-}NH_2}$$

$$\rightarrow \left[\mathrm{-OC{-}(CH_2)_4{-}CO{-}NH{-}(CH_2)_6{-}NH-}\right]_n$$

$$+(n-1)\mathrm{H_2O}$$

If you consider the examples given above you will see that only compounds with appropriate reactive groups can act as monomers.

What reactive group must a compound have so that it might be a monomer suitable for addition polymerization? As far as you can, make a list of plastics made by addition polymerization and draw the structural formulae of the corresponding monomers.

For condensation polymerization, how many reactive groups must there be in each molecule of monomer? Make a list of functional groups which occur in organic compounds, and then a further list of pairs which might be suitable for condensation polymerization.

Carothers' experimental work suggested that there are important relationships between the molecular shape and the physical properties of polymers.

Consider the following reactions:

1 Succinic acid, $HO_2C—CH_2—CH_2—CO_2H$, and ethane-1,2-diol, $HOCH_2—CH_2OH$, were heated at 160–175 °C and water was evolved. The temperature was raised to 250 °C to complete the reaction and on cooling a hard brittle opaque white mass was obtained. By recrystallization from ethyl acetate a microcrystalline powder was obtained, with melting point 102 °C, and molecular weight of about 3000.

2 Phthalic acid, $C_6H_4(CO_2H)_2$ (benzene ring with two adjacent CO_2H groups) and ethane-1,2-diol when similarly treated produced a hard transparent glass-like resin of molecular weight about 4800.

Which product has a well-ordered structure? Consider the molecular shapes of the two polymers. Can you see that one is more likely to pack easily into a well-ordered structure?

How many benzene dicarboxylic acids are there? Which is most likely to produce a crystalline polymer? Is this polymer known to you?

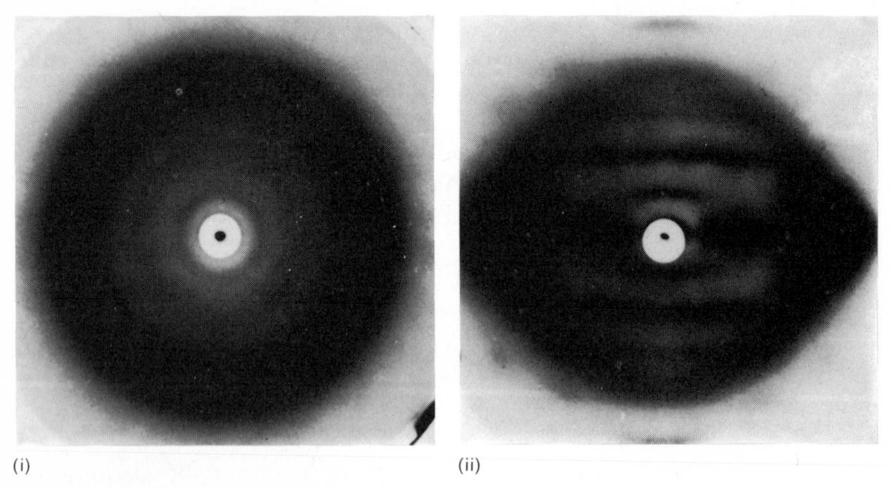

(i) (ii)

Figure 18.8a
X-ray fibre photograph of 15 denier nylon monofilament.
i undrawn monofilament.
ii cold drawn (× 5) monofilament.
Photo, ICI Fibres Ltd

Figure 18.8a shows X-ray fibre photographs of nylon drawn and undrawn, that is, before and after stretching. How well-ordered is the structure of undrawn nylon and how are the ordered regions affected by the drawing process? Is the molecular shape of nylon suitable for the formation of a well-ordered structure?

The 'drawing' of nylon (type 66) can be further investigated. Examine a narrow strip of nylon film about 25 cm long between crossed polaroids. Is the nylon film isotropic or anisotropic? Look again at section 8.2 for an explanation of these terms. Now grasp the nylon strip firmly and stretch it until it breaks, observing the film between crossed polaroids at stages during the process of stretching.

Make a careful record of your observations and compare your results with Carothers' original description of 'cold drawing'.

> In connection with the formation of fibres the polymers exhibit a rather spectacular phenomenon which we call cold drawing. If stress is gently applied to a cylindrical sample of the opaque, unoriented polymer at room temperature or at a slightly elevated temperature, instead of breaking apart, it separates into two sections joined by a thinner section of the transparent, oriented fibre. As pulling is continued this transparent section grows at the expense of the unoriented sections until the latter are completely exhausted. A remarkable feature of this phenomenon is the sharpness of the boundary at the junction between the transparent and the opaque sections of the filament. During the drawing operation the shape of this boundary does not change; it merely advances through the opaque sections until the latter are exhausted. This operation can be carried out very rapidly and smoothly, and it leads to oriented fibres of uniform cross section. The oriented and unoriented forms of the polymer are different crystalline states, and in the cold drawing operation one crystalline form is instantly transformed into the other merely by the action of very slight mechanical stress.

In short, (1) if a crystalline polymer is to be obtained, the long chain molecules of the polymer should have a good degree of molecular symmetry, and (2) if the crystalline regions of a polymer can be orientated in the same direction by 'cold drawing', a strong fibre is obtained.

The degree of crystallinity and orientation in a sample will depend on its treatment. For moulded samples of a given polymer the degree of crystallinity can be judged from the density: for example, for nylon (type 66) a density of 1.20 g cm^{-3} corresponds to 75 per cent crystalline, and 1.12 g cm^{-3} to 20 per cent crystalline; 40 per cent crystallinity is the degree usual in moulded articles.

Background reading

1 Nylon stockings

One of the most popular uses for nylon fibres is in stockings, and the following extract from Carothers' original patent (U.S. 2 157 116) applied for on 15th February, 1937, is included for those who may be especially interested:

> This invention has as an object the preparation from synthetic materials of knitted hosiery equal to silk hosiery in appearance and superior in elasticity and wearing qualities.
>
> This object is accomplished by preparing the knitted hosiery from synthetic linear condensation polyamides.
>
> In general, knitting gives a more elastic fabric than weaving, and for this and other reasons is used in the preparation of hosiery which must give a snug fit and which must at the same time be capable of considerable stretching without being permanently deformed.
>
> To prepare an elastic hose, threads or yarns must be used which can be set in a wavy shape by some process like boarding, and when so set the bent threads must tend to retain their shape with sufficient force to overcome the friction between the threads at all humidities encountered in wear.
>
> At the present time, the hosiery field is dominated by silk, since silk is the only fibrous material known which has the necessary elastic properties. Approximately 80 per cent of the silk used in this country goes into this outlet alone. Numerous attempts have been made to supplant silk in hosiery with artificial fibres, namely, those of the cellulose type, such as viscose rayon, but without success. Stockings prepared with these artificial yarns, owing to the poor bending elasticity of the fibres, have a greater tendency to wrinkle at the knee and ankle than silk stockings, a greater tendency to enlarge in diameter and shorten in length, and a tendency to slip over the skin in adjusting to movements of the leg instead of following the skin as does a silk stocking. In other words, they do not 'cling' properly and soon become 'baggy' at the knees. As a result, stockings prepared from these artificial fibres are inferior to silk stockings and cannot command so good a price.
>
> It has now been found that knitted wear, and particularly stockings, fully equal to those obtained from silk can be prepared from synthetic linear condensation polyamide yarns; in fact, most yarns derived from such polyamides have been found to yield knitted fabrics which are superior to silk. The synthetic polyamide stockings cling properly and do not become baggy.
>
> While the properties of the synthetic polyamide fibres will of course vary somewhat with the nature of the reactants used in their preparation, common characteristics of the fibres are high tenacity, high degree of fibre orientation, lack of sensitivity to conditions of humidity, extraordinary resistance to solvents and chemical reagents, exceptionally good elastic recovery, good dyeing properties, and good ageing characteristics, in air even at elevated temperatures. Although the fibres are resistant to most chemical reagents, they yield on hydrolysis with strong mineral acids the monomeric reactants from which they were derived.

A valuable property of the synthetic polyamides is that they can be spun from melt, that is, by extruding a molten polymer through suitable orifices and cold drawing the filaments thus prepared. This cold drawing (stretching in the solid state) causes fundamental physical changes to take place in the filament which then shows under X-ray examination orientation along the fibre axis and exhibits the physical properties most desired in fibres, as for instance great strength and pliability. Fibres prepared in this way resemble silk in certain respects; they are continuous, straight, and smooth, although fibres having crenulated surfaces can be prepared under suitable spinning conditions, e.g. by wet and dry spinning methods. While the fibres are, in general, lustrous, they can be delustered. Delustered yarns are generally preferred in the preparation of stockings, since this tends to 'slenderize' the ankles. If dyed fabrics are desired, the dye may be applied to the finished fabrics or to the yarn used in knitting. It may also be incorporated in the polyamide from which the yarn is prepared.

In the preparation of polyamide knitted wear it is generally desirable to use fibres which have been substantially completely cold drawn and to prepare the fabric in essentially the size and shape it is to assume in use. Such fibres usually have a residual elongation of 15 to 40 per cent and have good elasticity. They require a greater force to elongate them than undrawn or partially drawn cold fibre. It is within the scope of this invention, to knit fabrics from undrawn or partially drawn fibres. These fibres have a very high residual elongation, often as much as several hundred per cent. When such fibres are subjected to stress, they are capable of undergoing considerable elongation without breaking. This is sometimes desirable, e.g., in the preparation of a stocking, since it gives a product which is not easily broken. For example, when the stocking becomes caught on a protruding object, the wearer becomes aware of the fact and can release the stocking before it breaks. In view of the surprisingly good elasticity and wearing qualities of the synthetic linear condensation polyamide knitted fabrics, the products of this invention represent a phenomenal advance in the textile art. In many respects, these fabrics are as much superior to silk fabrics as the latter are superior to fabrics made from cellulose materials. These fabrics are fully equal to silk in strength and are markedly superior in elastic behaviour particularly under high degrees of stretch and at high humidities. Other advantages of polyamide stockings over silk stockings are that they are less sensitive to deterioration at elevated temperatures and less easily spotted by water. Furthermore, the stockings of this invention are essentially creaseproof under ordinary conditions but retain creases introduced with steam. This invention, therefore, represents the first truly successful stocking from a synthetic yarn.

2 Polythene – a chance discovery

Ethylene was first polymerized in 1933 by R. O. Gibson and E. W. Fawcett while working in the Winnington Hall laboratories of ICI. However it was not until the work was repeated eighteen months later by M. W. Perrin that sufficient high polymer was obtained for a technical evaluation to be carried out, and for the importance of the discovery to be appreciated.

The story of the discovery is a good example of how unexpected experimental results need to be carefully considered, for they may be quite as valuable as the result that was hoped for.

High pressure studies had been started by ICI in connection with the Haber process but in 1931, in consultation with Sir Robert Robinson, then professor of organic chemistry at Oxford, attention was directed at 'the possibility of bringing about organic chemical reactions, particularly those of condensation type, which at lower pressures require the aid of catalysts'. It was suggested that work should be carried out with simple unsaturated compounds whose reactivity it was hoped to enhance by application of pressures greater than 1000 atmospheres and Gibson and Fawcett reported in April 1933:

Ethylene and Benzaldehyde

Attempts have been made to cause the following reaction to occur:

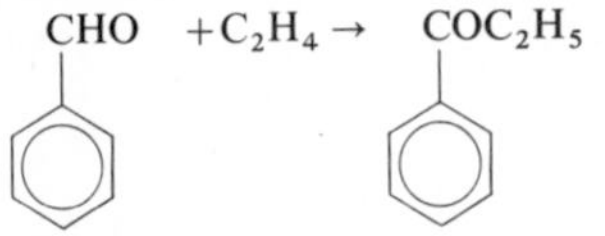

$$C_6H_5CHO + C_2H_4 \rightarrow C_6H_5COC_2H_5$$

At 170 °C and 2000 atm pressure, reaction occurred slowly to yield a hard waxy solid containing no oxygen, melting at 113 °C and analysing to $(CH_2)_n$ – apparently an ethylene polymer. In some cases during the investigation of this reaction, a violent reaction has occurred with considerable rise in pressure and on opening up the bomb it has been found that complete carbonization of the charge has taken place. In these experiments (i.e. in which carbonization has occurred), simple decomposition of ethylene has apparently taken place.

By July about half a gramme of polymer had been obtained in all, but the explosive carbonization could not be controlled and work ceased.

To pause on the brink of an important discovery may seem strange but it should be remembered that in 1933 the common plastics (such as phenol-formaldehyde) were rigid materials while the polyethylene obtained was soft, and the quantity available was insufficient for technical evaluation. Nor was polymerization the object of the research (unlike Carothers' work) so the high pressure studies were continued using carbon monoxide instead of the explosive ethylene. However in 1935 Perrin looked again at the polymerization of ethylene and by good fortune obtained sufficient yields for the importance of the discovery to be realized:

December 1935

In one experiment at 2000 atm and 170 °C where the gas was compressed directly into a steel bomb, the reaction proceeded steadily to give about 8 g in four hours of a white waxy solid polymer of ethylene. The molecular weight of the polymer is about 3000. The properties of this substance are being studied. Other experiments have given an explosive decomposition of the ethylene to carbon and hydrogen.

January 1936

Further quantities of the polymer have been made. Some of this has been moulded at about 150 °C under 200 atm pressure. A small plasticizing mould and die have been made for this work. A preliminary figure shows that the electrical volume resistivity of the moulded polymer is greater than 5×10^{14} ohm cm^{-3} showing that it falls into the class of good dielectrics.

It has been possible to make thin transparent films of the polymer which have considerable strength and toughness.

It is noteworthy that polytetrafluoroethylene (PTFE) came about by a similar chance discovery. While working for Kinetic Chemicals Inc. on refrigerant gases Dr Roy Plunkett noticed that a cylinder of tetrafluoroethylene gas displayed no gas pressure but was still at its full weight. On cutting open the cylinder a white solid was found which proved to be polymerized tetrafluoroethylene.

Polythene film can be tested for elasticity and the ability to be cold drawn in the same way as nylon. Strips of film (2×10 cm) should be cut with a sharp knife otherwise the film may tear rather than stretch when under stress. Furthermore it is worth while cutting two strips, one parallel to the edge of the film and the other at right-angles to see if the manufacturing process results in any orientation of crystalline regions in the film.

Examine the strips of film between crossed polaroids and pull gently so as to stretch the polythene. Can polythene be cold drawn and does cold drawing result in a gain in strength? Are the two strips of film identical in behaviour?

Low density polythene (the form obtained by the high pressure process outlined above) is about 60 per cent crystalline; it has been suggested that the initial stretch is taken up in the amorphous parts of the structure, and in the subsequent cold drawing process the crystalline regions become orientated along the 'fibre axis', with a resultant increase in strength.

The X-ray powder photograph of low density polythene (figure 18.8b) indicates its crystalline nature and the well orientated crystalline nature of PTFE is indicated by its X-ray fibre photograph (figure 18.8c).

The molecular shapes of polythene and PTFE can be deduced from their X-ray photographs and are illustrated in the same figures.

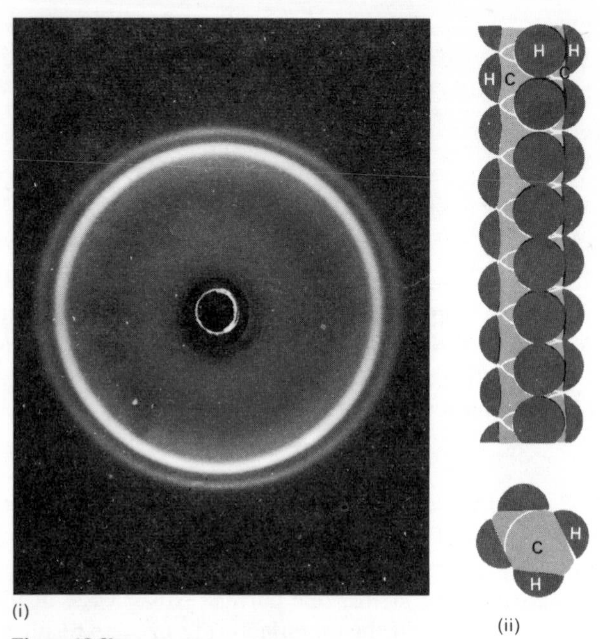

Figure 18.8b
i X-ray powder photograph of L.D. polythene.
ii Structure of the polythene molecule.

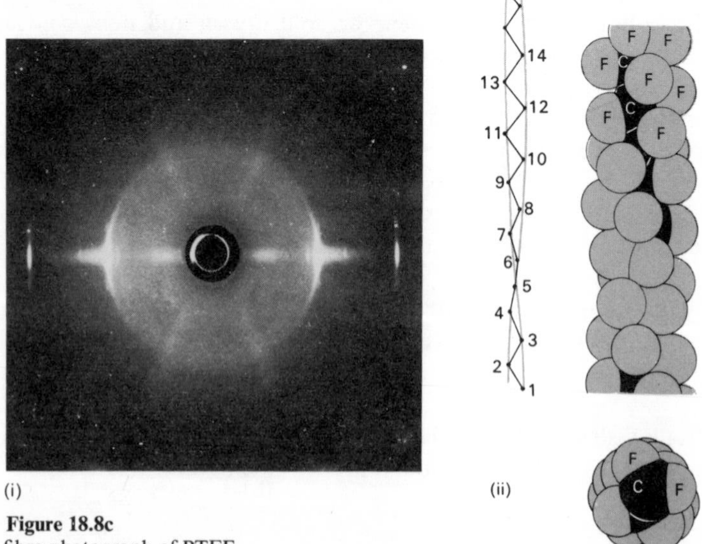

Figure 18.8c
i X-ray fibre photograph of PTFE.
ii Structure of the molecule of PTFE; conversion of a polyethylene chain to a helix.
Figures 18.8b and 18.8c: photos, ICI Plastics Ltd; drawings after Bunn, C. W. and Howells, E. R. (1954) Nature, **174**, *549*

The carbon-carbon spacing along the axis of an alkane is 0.127 nm, so hydrogen atoms can fit above one another along the molecule because they require a space of radius only 0.120 nm. In this case, the van der Waals' forces between the non-bonding atoms are acceptable. But a fluorine atom requires a space of radius 0.140 nm, so in the case of PTFE the molecule adopts a twisted chain resulting in the fluorine atoms forming a helix on the surface of the carbon chain. The X-ray data indicate that there is a regular repeat every 1.68 nm along the molecular axis, that is every (1.68/0.127) atoms or 13 atoms. It has been proposed that the very smooth surface profile of the PTFE molecule allows the molecules to slip readily over one another and this results in the remarkably low coefficient of friction (0.02–0.1).

The work of Ziegler and Natta on polymerization catalysts

Professor Karl Ziegler, working at the Max Planck Institute for Fuel Technology in 1949, observed a 'growth reaction' between aluminium alkyl compounds and ethylene which produced useful yields of C_6 to C_{10} compounds:

$$Al(-CH_2-CH_3)_3 + nCH_2{=}CH_2 \rightarrow \text{/\/\/\/}-Al(\text{/\/\/\/})_2$$

where /\/\/\/— represents an alkyl chain.

Attempts were made to use this growth reaction to produce straight-chain polymers of high molecular weights. These attempts failed because the reaction rate was slow and a side reaction, which became more pronounced if the temperature was raised, prevented the attainment of satisfactory polymers;

$$\text{/\/\/\/}-Al(\text{/\/\/\/})_2 + 3CH_2{=}CH_2 \rightarrow Al(-CH_2-CH_3)_3 + 3\,\text{/\/\/\/}$$

A chance discovery led to the hypothesis that minute traces of metal salts caused the side reaction. Research was therefore undertaken to find which salts might be responsible and hence devise a purification procedure to eliminate them from the starting materials. With the total elimination of the offending metal salts it was hoped that the growth reactions would produce useful polymers.

In late 1953 a zirconium salt was tested to see if it caused the side reaction. This was not so. Instead the zirconium salt proved to be an excellent catalyst for the growth reaction giving a high molecular weight polyethylene as product. Furthermore the polymerization proceeded, not at 2000 atmospheres and 200 °C as in the ICI process, but at normal pressure and without external heating.

At this time Giulio Natta, working in the Polytechnic Institute of Milan, learnt of Ziegler's results and studied the polymerization of other alkenes by the same technique. Propylene, $CH_3—CH{=}CH_2$, was a starting material of interest because it was available in quantity from petrochemical processes and high-pressure polymerization had yielded only worthless gums. In early 1954 Natta found an aluminium triethyl–titanium tetrachloride combination which would successfully polymerize propylene.

Write down the structural formulae of propylene and polypropylene (three or four monomer molecules only) and note whether the polymerization is addition or condensation.

Natta examined his polypropylene product with care and was able to divide it into fractions by solvent extraction. He found two main fractions.

1 An amorphous fraction of low molecular weight. What type of molecular shape would you predict for an amorphous material? Such polymers are known as *atactic* (a-tactic: dis-order).

2 A crystalline fraction that ranged from low to high molecular weight. What type of molecular shape would you predict for a crystalline material? X-ray fibre photographs indicate that the monomer units are joined head-to-tail in a regular way giving a repeating pattern. Such polymer structures are known as *isotactic* (iso-tactic: same-order). The isotactic structure for polypropylene is illustrated in schematic form in figure 18.8d. This structure gives the best balance between the interatomic forces of repulsion and the van der Waals' forces of attraction between the methyl groups.

n $CH_2{=}CH{-}CH_3$ (monomer) → $[CH_2{-}CH(CH_3){-}CH_2{-}CH(CH_3){-}\ldots]_n$ (isotactic polymer)

CH2
n CH
CH3
monomer

CH2 CH2
CH CH
CH3 CH3
isotactic polymer

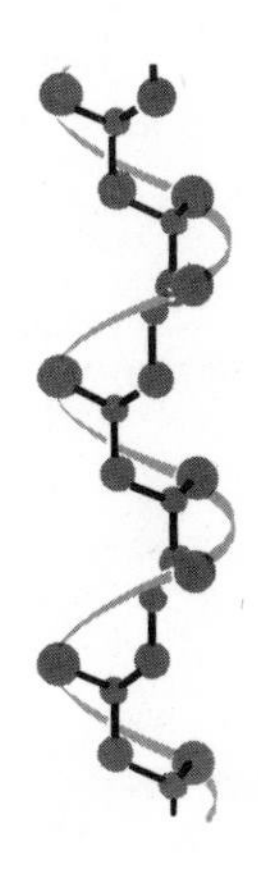

Isotactic chain making one complete
turn for every 3 monomer units

Figure 18.8d
The structure of isotactic polymers.

Another type of regular structure was noted by Natta as a polymerization product from butadiene, $CH_2{=}CH{-}CH{=}CH_2$. This structure is known as *syndiotactic* (syn-di-o-tactic: together-two-order) and the monomer units are joined in repeating pairs. The syndiotactic structure of polypropylene is illustrated in figure 18.8e.

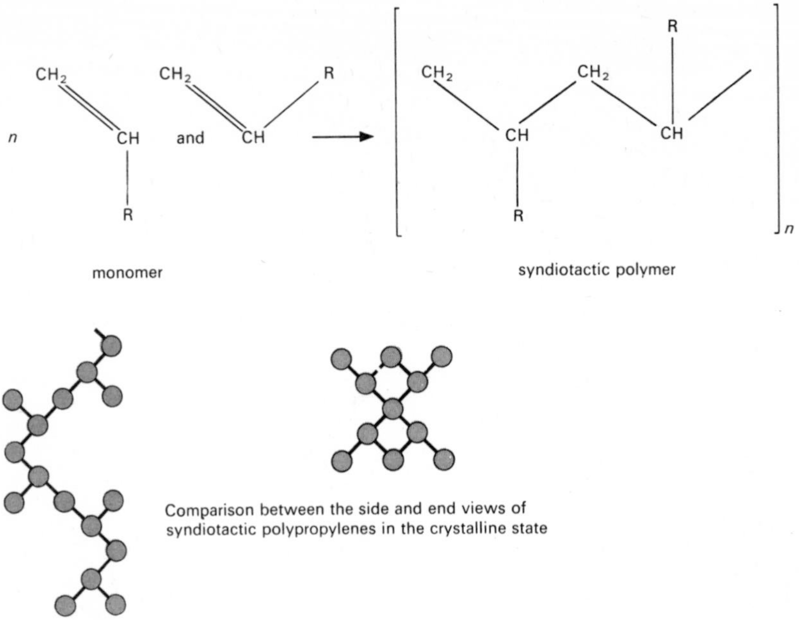

monomer

syndiotactic polymer

Comparison between the side and end views of syndiotactic polypropylenes in the crystalline state

Figure 18.8e
The structure of syndiotactic polymers.

These polymers of regular molecular shape obtained by the use of Ziegler catalysts are known as *stereoregular*. They have significant technical differences from the irregular atactic polymers. Thus the relative degree of crystallinity in low density polythene (obtained by the high pressure process) and high density polythene (obtained by the use of a Ziegler catalyst) can be judged from a comparison of their X-ray powder photographs (figures 18.8b and 18.8f).

The change in properties due to a more ordered structure is indicated in table 18.8.

	L.D. polythene	**H.D. polythene**
Density/g cm^{-3}	0.92	0.96
Tensile strength/N m^{-2}	10×10^6	30×10^6
Approximate elongation at break/per cent	400	200
Melting range/°C	110–115	130–135

Table 18.8
The properties of L.D. and H.D. polythene

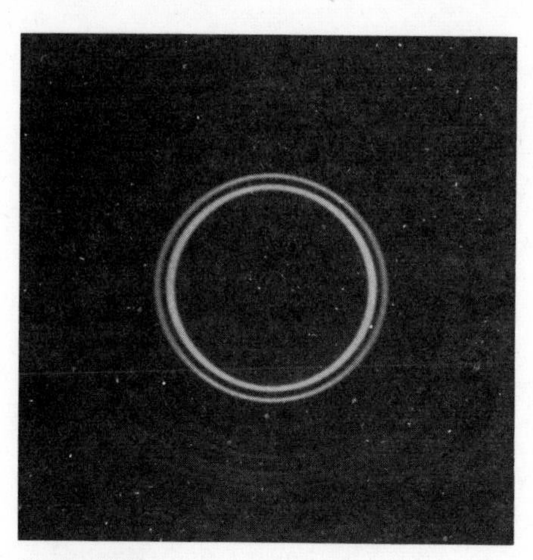

Figure 18.8f
X-ray powder photograph of H.D. polythene.
Photo, ICI Plastics Ltd

Quite remarkable results have been obtained by the stereoregular polymerization of isoprene. Two products have been obtained:

1 A catalyst system of aluminium triethyl (1 mole) and titanium(IV) chloride (1 mole) produces an elastic polymer, density 0.92 g cm^{-3}, melting point 28 °C, X-ray fibre repeat 0.81 nm, identical with pure natural rubber.

2 A catalyst system of aluminium triethyl (1 mole) and titanium(IV) chloride (3 moles) produces a hard inelastic polymer, density 0.95 g cm^{-3}, melting point 65 °C, X-ray fibre repeat 0.88 nm, identical with gutta-percha.

Golf balls are made with an outer cover which is 90 per cent gutta-percha and an inner winding of natural rubber. Their relative properties can be studied by cutting up an old golf ball.

The differences in properties are due to a change in molecular structure; natural rubber is *cis*-1,4-polyisoprene and gutta-percha is *trans*-1,4,-polyisoprene (figure 18.8g).

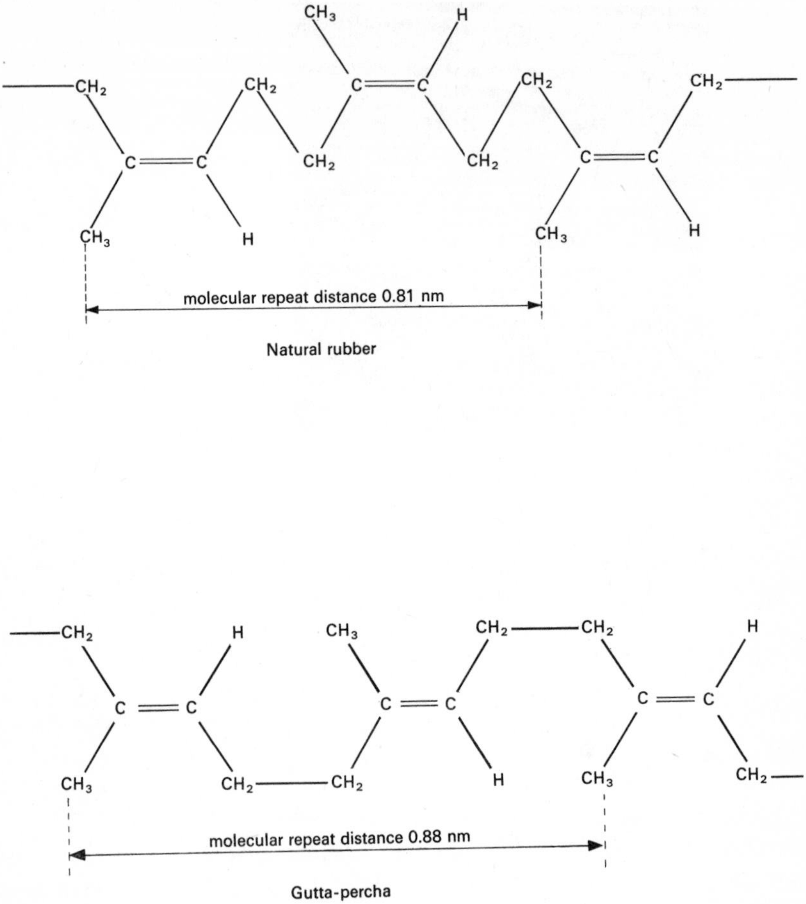

molecular repeat distance 0.81 nm

Natural rubber

molecular repeat distance 0.88 nm

Gutta-percha

Figure 18.8g
The molecular structure of rubber and gutta-percha.

Topic 19

Some p-block elements

19.1 **Nitrogen**

Nitrogen can have oxidation numbers from -3 to $+5$.

$+5$ —	NO_3^-	(nitrate ion)	
$+4$ —	N_2O_4	(dinitrogen tetroxide) $\rightleftharpoons$ NO_2 (nitrogen dioxide)	
$+3$ —	NO_2^-	(nitrite ion)	
$+2$ —	NO	(nitrogen monoxide)	
$+1$ —	N_2O	(dinitrogen monoxide)	
0 —	N_2		
-1 —	NH_2OH	(hydroxylamine)	
-2 —	N_2H_4	(hydrazine)	
-3 —	NH_3	(ammonia)	NH_4^+ (ammonium ion)

Figure 19.1a
Oxidation states of nitrogen.

The important oxidation numbers are $+5$, $+4$, $+3$, 0, and -3.

One way of thinking about the compounds that nitrogen forms is to look at experimentally determined ionization energies. These, of course, are the energies required to remove electrons completely from the atom, but they will also reflect the *availability* of electrons for chemical reactions. If we look at successive ionization energies for elements across the Periodic Table from lithium to fluorine, we notice a sudden increase in the energy when we break into the 'helium core' of electrons (see figure 19.1b).

$$\begin{aligned} He &= 1s^2 \\ Li &= \text{'He core'} + 2s^1 \\ Be &= \qquad + 2s^2 \\ B &= \qquad + 2s^2\,2p^1 \\ C &= \qquad + 2s^2\,2p^2 \\ N &= \qquad + 2s^2\,2p^3 \\ O &= \qquad + 2s^2\,2p^4 \\ F &= \qquad + 2s^2\,2p^5 \end{aligned}$$

that is, when we form Li^{2+}, Be^{3+}, etc. This indicates that one electron in the case of lithium, two in the case of beryllium, etc. are available for bonding. In the case of lithium, the charge on the ion is small so the simple ion, Li^+, is formed. In other cases, the charge on a simple ion like this would be too large and a covalently bonded ion or molecule is formed in which, in the case of carbon, four electrons are used for bonding.

Five electrons are available in the nitrogen atom without breaking into the 'helium core'. The electrons are arranged in energy levels as shown in figure 19.1c.

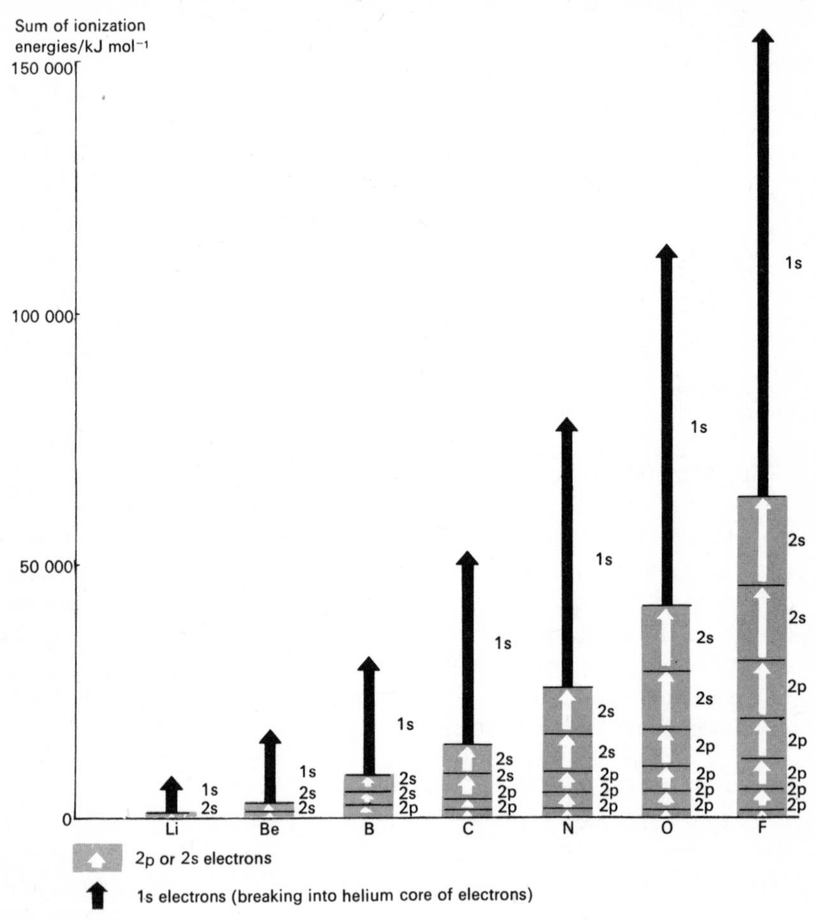

Figure 19.1b
Ionization energies, lithium to fluorine.

Except for those in the 2p level, the electrons are *paired*. Electron pairing often occurs, as when a covalent bond is formed. In the atom, if it is possible, electrons do not pair – this is seen in the 2p level of nitrogen, but, as in the 1s and 2s levels, some pairing must occur.

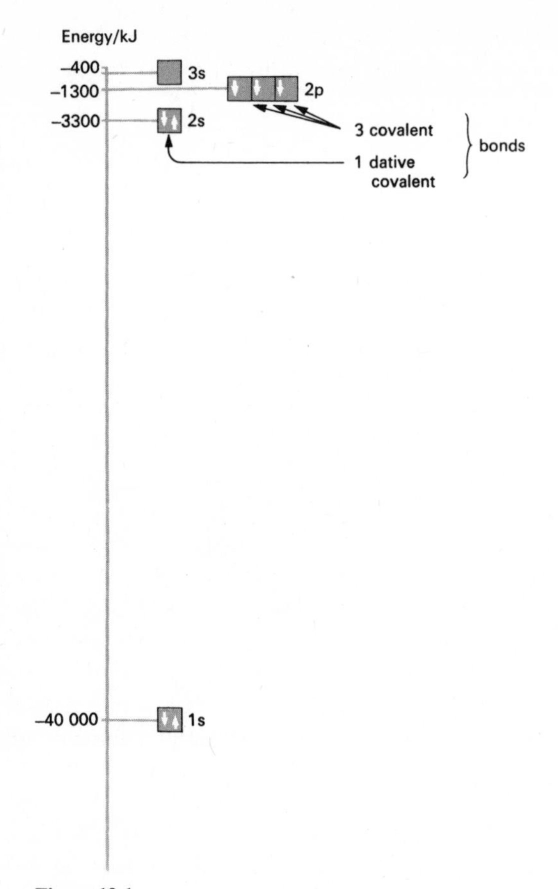

Figure 19.1c

The five electrons that are available must be those in the 2s and 2p levels and and it can be seen that three 'normal' covalent bonds and one dative covalent bond can be formed. Nitrogen has its highest oxidation number in the nitrate ion so the highest number of electrons from the nitrogen atom, five, is probably involved (oxygen atoms all having the electronic configuration of neon):

$$\left[\begin{array}{ccc} \mathrm{O} & & \mathrm{O} \\ & \mathrm{N} & \\ & \mathrm{O} & \end{array}\right]^{-} \quad \text{or} \quad \left[\begin{array}{ccc} \mathrm{O} & & \mathrm{O} \\ & \diagdown \quad /\!\!/ & \\ & \mathrm{N} & \\ & | & \\ & \mathrm{O} & \end{array}\right]^{-}$$

As we know that the bond lengths in the nitrate ion are all the same, the formal double bond we have indicated does not occur: the electrons are delocalized:

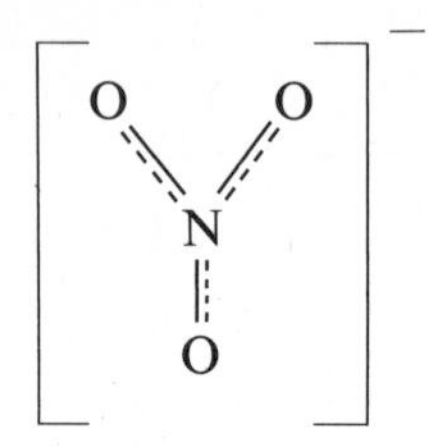

If we apply oxidation number rules to the nitrate ion, we find that the oxidation number of nitrogen in it is +5. *In this case*, the oxidation number coincides with the number of electrons used for bonding.

We have found that the oxidation number is a very useful concept in dealing with redox reactions. *In many cases*, it also coincides with the number of electrons used, and as long as there is evidence that this is so, it is useful to think of the oxidation numbers of elements which have variable oxidation numbers in this way. In the nitrite ion and in ammonia in which the oxidation numbers of nitrogen are respectively +3 and −3, the number of electrons used is three. In other cases, such as when there are two or more atoms of the element in question in the molecule or ion such as in N_2O_4 or N_2O, or when the oxidation number is zero as in N_2, there is no correlation between the oxidation number and the number of electrons used for bonding.

A possible way in which nitrogen could form five 'normal' covalent bonds is for one of the electrons in the 2s level to be promoted to the next available level, 3s, so giving five unpaired electrons. This does not happen because the energy required is about 3000 kJ and the energy obtained from the formation of the two covalent bonds (the energy of the N—O bond in nitrogen monoxide is about 630 kJ) does not provide sufficient compensation.

For phosphorus, a 3d level is available, only 1370 kJ above the 3s, and this promotion does occur, so we find that PCl_5 as well as PCl_3 occurs, whilst nitrogen only forms one chloride, NCl_3. (See figure 19.1e.)

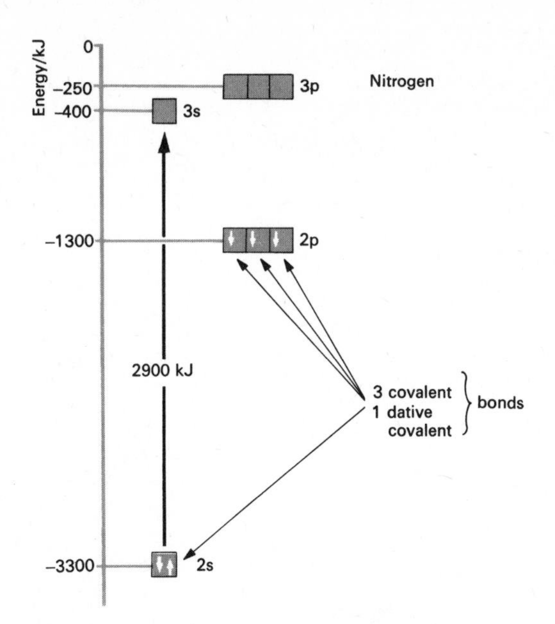

Both nitrogen and phosphorus are essential to life: the adenosine phosphates which function as energy carriers in living organisms, and the nucleic acids, contain phosphorus. Proteins, the important constituents of plant and animal tissue, contain about 16 per cent nitrogen. Nearly 80 per cent of the atmosphere is nitrogen, but the conversion of this to compounds of nitrogen which plants can use – *the fixation of nitrogen* – is difficult because of the high stability of the nitrogen molecule.

This molecule has a triple bond:

$$:N \overset{\times}{\underset{\times}{\dot{\times}}} N \overset{\times}{\times} \qquad \text{or} \qquad :N \equiv N:$$

(*Note*. Oxidation number of nitrogen = 0 but three electrons are used for bonding.)

The bond dissociation energy of N_2 is high, as is that for other triple bonds such as that in acetylene, $HC \equiv CH$:

$N \equiv N$ 946 kJ mol^{-1}
$HC \equiv CH$ 954 kJ mol^{-1}
Compare $H—CH_3$ 431 kJ mol^{-1}

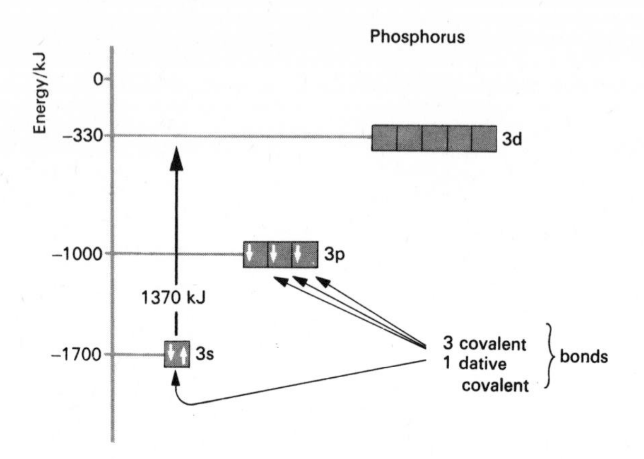

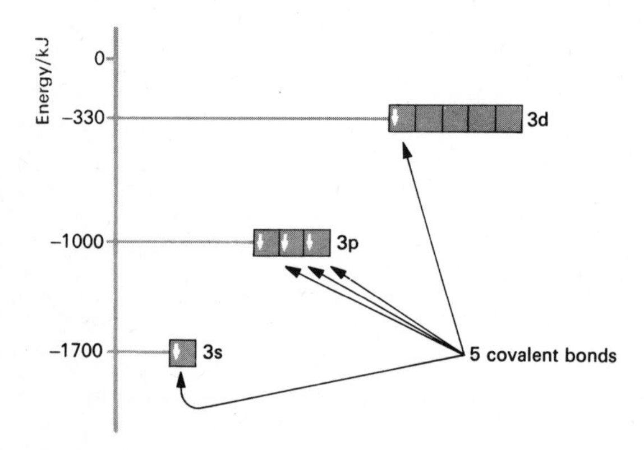

Figure 19.1e

In acetylene, four of the electrons are available for addition reactions similar to those of ethylene, $CH_2{=}CH_2$. For instance:

$$HC{\equiv}CH + 2H_2 \rightarrow CH_3{-}CH_3$$

But the electrons in $N{\equiv}N$ are not similarly available. This is shown by the ionization energy of the nitrogen molecule

$$N_2 \rightarrow N_2^+ + e^-; \quad IE = 1506\ \text{kJ mol}^{-1}*$$

as compared with

$$C_2H_2 \rightarrow C_2H_2^+ + e^-; \quad IE = 1110\ \text{kJ mol}^{-1}$$

* Note that the ionization energy normally quoted for nitrogen is for the atom $N \rightarrow N^+ + e^-; IE = 1402\ \text{kJ mol}^{-1}$

1506 kJ mol^{-1} is nearly the ionization energy of the inert gas, argon:

$$Ar \rightarrow Ar^{+} + e^{-};\ IE = 1523\ \text{kJ mol}^{-1}$$

indicating how difficult it is to use the electrons in the nitrogen triple bond. This stability of nitrogen accounts for the positive heats of formation of the nitrogen oxides:

Oxide	$\Delta H_f^{\ominus}$**/kJ mol^{-1}**	$\Delta G_f^{\ominus}$**/kJ mol^{-1}**
$N_2O(g)$	+82.0	+104.2
$NO(g)$	+90.4	+86.6
$NO_2(g)$	+33.2	+51.3
$N_2O_4(g)$	+9.2	+97.8

What nitrogen compounds can be made from nitric acid?

As the oxidation number of nitrogen will change, redox potentials should help us. As we found in Topic 16 (Some d-block elements) electrode potentials can be used to predict whether a reaction is possible under standard conditions. Figures 19.1f and 19.1g give $E^{\ominus}$ values for various nitrogen compounds at pH = 0 and pH = 14.

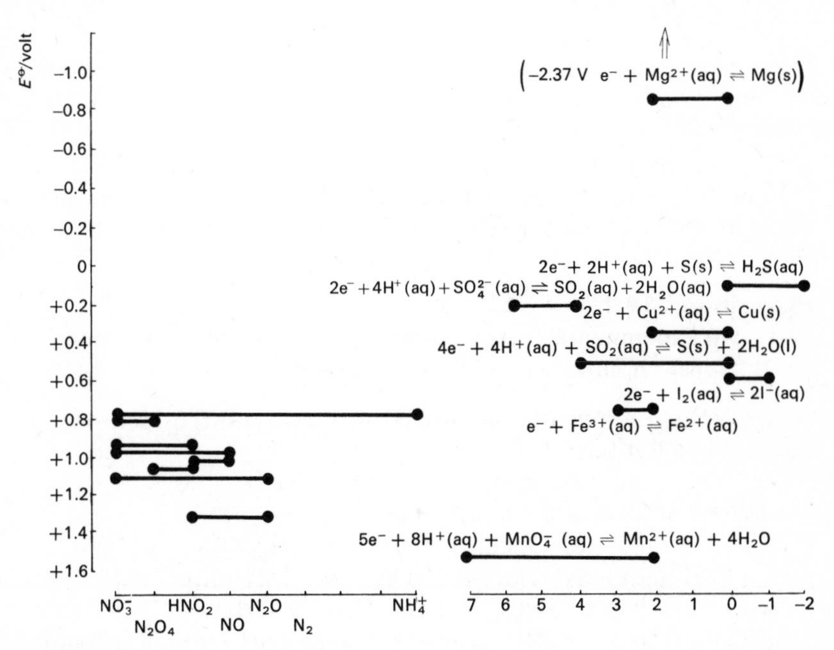

Figure 19.1f
Redox potentials at pH = 0.

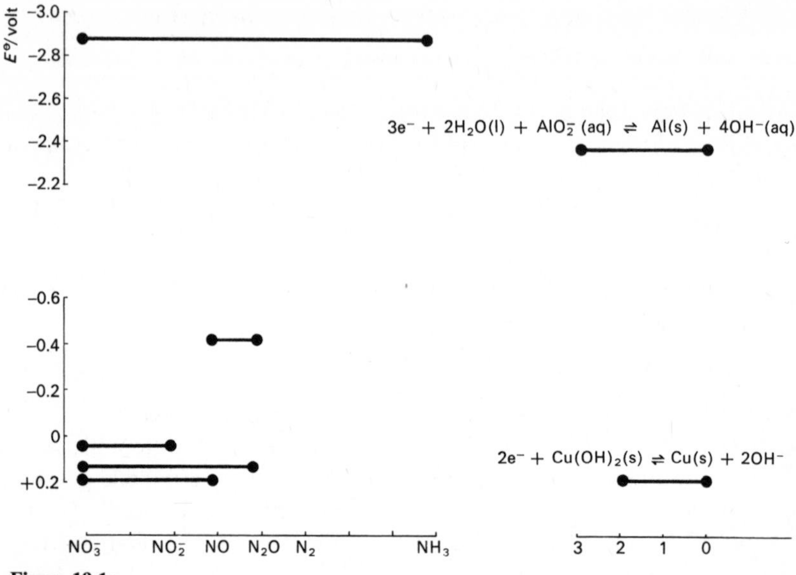

Figure 19.1g
Redox potentials at pH = 14.

These charts only indicate that a reaction is possible. They tell us nothing about how quickly or slowly the reaction proceeds.

We can see from the charts that common reducing agents, like Mg, H_2S, Cu, I^-, and Fe^{2+} theoretically could reduce NO_3^- to NH_4^+. But the reaction may be controlled kinetically and stop at some other oxidation number of nitrogen.

Experiment 19.1

An experimental investigation of the chemistry of nitrogen

What happens if nitric acid is added to copper?

a Add *a few drops* of concentrated nitric acid (14M) to *three or four* copper turnings in a test-tube.

1 What nitrogen compound has been formed?

Let us look at some of the possibilities:

$3Cu(s) + 12H^+(aq) + 6NO_3^-(aq) \rightarrow 3N_2O_4(g) + 6H_2O(l) + 3Cu^{2+}(aq)$

$3Cu(s) + 8H^+(aq) + 2NO_3^-(aq) \rightarrow 2NO(g) + 4H_2O(l) + 3Cu^{2+}(aq)$

$3Cu(s) + 7.5H^+(aq) + 1.5NO_3^- \rightarrow 0.75N_2O(g) + 3.75H_2O(l) + 3Cu^{2+}(aq)$

$3Cu(s) + 7.2H^+(aq) + 1.2NO_3^-(aq) \rightarrow 0.6N_2(g) + 3.6H_2O(l) + 3Cu^{2+}(aq)$

$3Cu(s) + 7.5H^+(aq) + 0.75NO_3^-(aq) \rightarrow 0.75NH_4^+(aq) + 2.25H_2O(l) + 3Cu^{2+}(aq)$

You have already seen what nitrogen compound is formed with concentrated acid,

2 What compound(s) might be formed with 7M nitric acid?

3 What compound(s) might be formed with 2M nitric acid?

b From concentrated (14M) nitric acid, carefully prepare about 4 cm^3 of approximately 7M acid in a test-tube and add to a spatula full of copper turnings. Wait and then compare the gas produced with that produced in (a).

4 What compound(s) have been formed?

5 Why can we not obtain ammonia from nitric acid under acid conditions?

c Heat about 2 cm^3 of dilute nitric acid with some Devarda's alloy and about 5 cm^3 of sodium hydroxide solution.

6 What gases are evolved? (There are two to be identified.)

Either or both of the two metals in Devarda's alloy could reduce the nitrate:

$$2OH^-(aq) + Cu(s) \rightarrow Cu(OH)_2(s) + 2e^-$$

$$4OH^-(aq) + Al(s) \rightarrow AlO_2^-(aq) + 2H_2O(l) + 3e^-$$

(aluminate ion)

7 Which metal is more likely to reduce nitrate ions?

8 What will be the effect of increasing the hydroxide ion concentration? (Work out the equation.)

d Test your prediction using copper and aluminium separately.

e Add concentrated nitric acid to some aqueous hydrogen sulphide solution and warm the mixture.

9 What do you think is the solid that is formed?

10 What do you think is the final oxidation number of the sulphur?

f Add concentrated nitric acid to an aqueous solution of iodide ions.

11 What is the change in the oxidation number of the iodine?

g From concentrated (18M) sulphuric acid, carefully prepare about 12 cm^3 of 6M acid by *adding acid to water*. (It is most important that the acid should be added to the water, and not the other way round. Serious accidents may be caused by acid being splashed if this precaution is not heeded.) Chill the diluted acid in an ice bath for at least five minutes, then add 0.5 g potassium or sodium nitrite *in very small portions with stirring*.

Warm 2 cm^3 of the solution.

12 What gas is evolved?

13 If part of the nitrous acid has been oxidized to nitrate, what has been reduced to what? (Look on $E^{\ominus}$ chart).

14 Is this possible in alkaline solution?

15 Is a similar reaction possible with nitrate?

16 How would you expect nitrous acid to react with
(i) permanganate ions, (ii) iodide ions?

h Test your predictions.

j Separately heat sodium (or potassium) and copper nitrates, identify the gases evolved and, after cooling, add dilute sulphuric acid to the residues.

17 What solids have been formed?

k Find the pH of solutions of potassium chloride and ammonium chloride.

18 Can you explain your observations?

19 What do you think will happen if magnesium powder is added to ammonium chloride solution?

l Test your prediction. (Warm the mixture if necessary.)

m Find the pH of 0.2M solutions of sodium hydroxide and ammonia.

20 What is the value of $[OH^-(aq)]$ in these solutions?

21 Why are they different?

n Complete the table below after adding the sodium hydroxide and ammonia solutions drop by drop, to about six drops of the metal ion solutions. Shake the solution after each addition and make sure that you add a large excess of alkali solution in each case.

	+0.2M NaOH(aq)	**+0.2M NH_3(aq)**
0.2M Pb^{2+}(aq)		
Ca^{2+}(aq)		
Zn^{2+}(aq)		
Cu^{2+}(aq)		

22 Account for your observations.

p Add about 2 cm^3 of a 0.2M ammonia solution to about 1 cm^3 of a 0.2M solution of magnesium ions. Add about 0.5 g ammonium chloride to about 2 cm^3 0.2M ammonia solution, then add this solution to a 0.2M solution of magnesium ions.

23 What happens to $[OH^-(aq)]$ and $[NH_3(aq)]$ in the ammonia solution when the ammonium chloride is added to it?

24 Account for the results in (p).

19.2 **Sulphur**

Sulphur appears in the same group as oxygen in the Periodic Table, and so one would expect that it would have the same number of electrons available for bonding as oxygen, that is 6; and this is so. A look at the ionization energies in figure 19.2a confirms this.

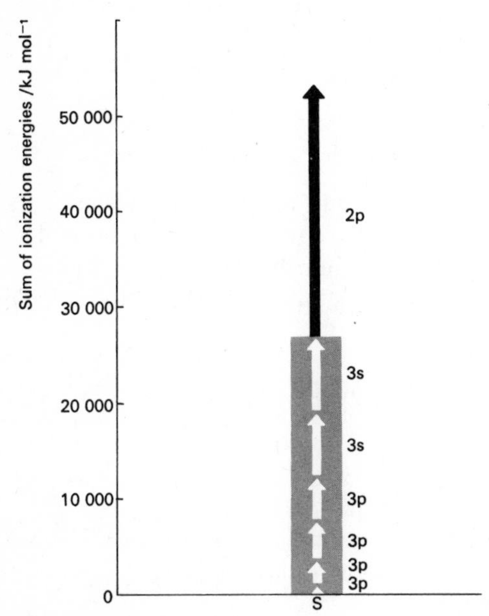

Figure 19.2a
Ionization energies of sulphur.

By similar considerations to those for nitrogen, it would be expected that electrons in oxygen could not be promoted from the second to the third energy level, but that electrons could be promoted within the third energy level from 3s and 3p to 3d in sulphur.

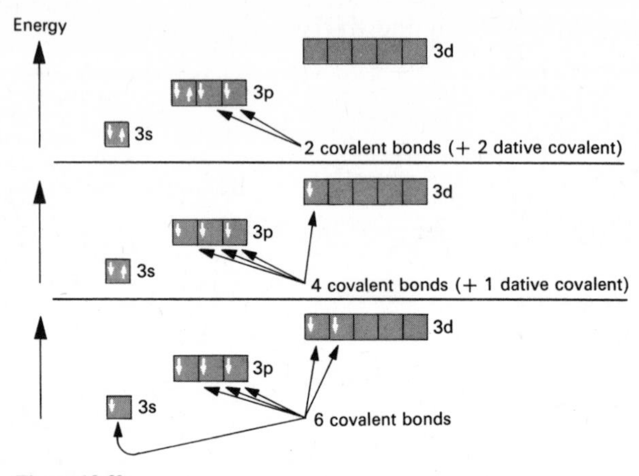

Figure 19.2b

This means that sulphur can form two, four, or six covalent bonds, whereas oxygen can form only two. Formation of two covalent bonds occurs in hydrogen sulphide (oxidation number −2) and sulphur dichloride (+2), four bonds in the sulphite ion and sulphur dioxide (+4), and six bonds in the sulphate ion and sulphur trioxide (+6).

Sulphur can have oxidation numbers ranging from +7 to −2 as shown in figure 19.2c.

[+7 —	persulphate,	$S_2O_8^{2-}$]	
+6 —	sulphate,	SO_4^{2-}	sulphur trioxide, SO_3.
+5 —	dithionate,	$S_2O_6^{2-}$	
+4 —	sulphite,	SO_3^{2-}	sulphur dioxide, SO_2.
+3 —	dithionite,	$S_2O_4^{2-}$	
+2 —	thiosulphate,	$S_2O_3^{2-}$	sulphur dichloride, SCl_2.
0 —	sulphur,	S_8	
−2 —	sulphide,	S^{2-}	hydrogen sulphide, H_2S.

Figure 19.2c

The electrons are delocalized in sulphate and sulphite ions in a similar manner to those in nitrate and nitrite ions (figure 19.2d).

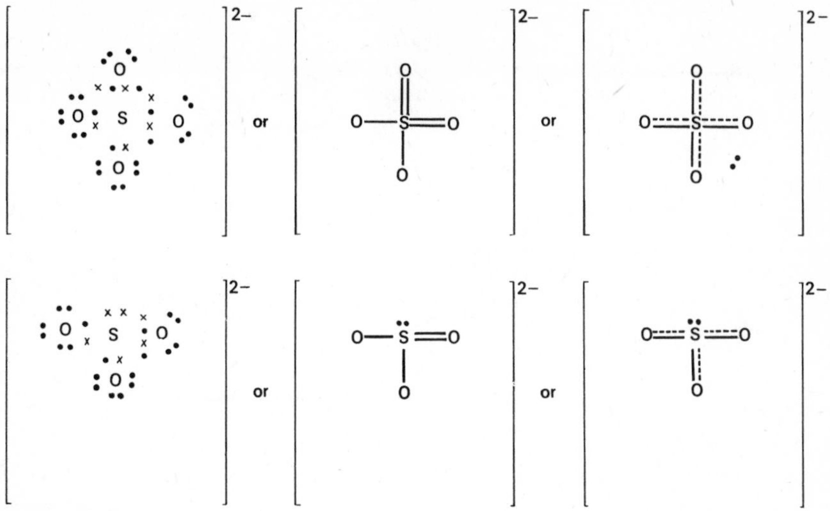

Figure 19.2d

We have seen that when there is a nitrogen–nitrogen bond, the oxidation number does not correspond to the number of electrons used, though it is still used in redox book-keeping. This is also true for compounds which contain sulphur–sulphur bonds:

Thiosulphate ion, $S_2O_3^{2-}$, $+2$
Tetrathionate ion, $S_4O_6^{2-}$, $+2\frac{1}{2}$ (the ion to which the thiosulphate ion is oxidized when it reacts with iodine)
Dithionite ion, $S_2O_4^{2-}$, $+3$
Dithionate ion, $S_2O_6^{2-}$, $+5$

The persulphate (or peroxydisulphate) ion, $S_2O_8^{2-}$, contains sulphur with an oxidation number of $+7$, which, on reduction to sulphate, changes to $+6$.

Assigning an oxidation number of $+7$ to sulphur in persulphate, works in redox book-keeping, but so would assigning an oxidation number of $+6$ to the sulphur and -1 to two of the oxygen atoms, which changes to -2 on reduction to sulphate. This unusual oxidation number of oxygen comes about because there is an oxygen–oxygen bond in persulphate, the structure of which is shown in figure 19.2e.

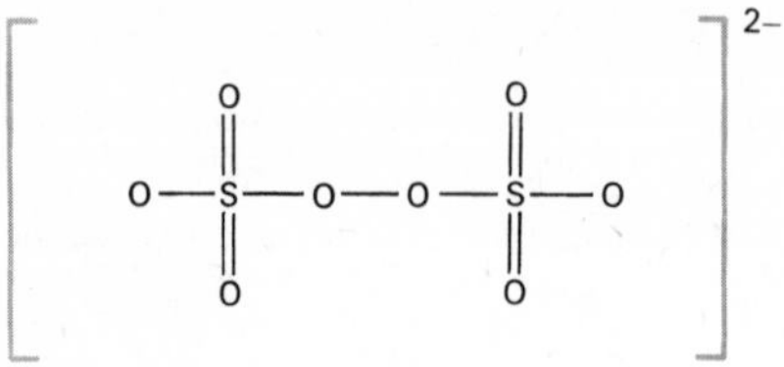

Figure 19.2e

You should now look at some of the reactions of some of these compounds experimentally. The compounds to be used are set out on the oxidation number chart (figure 19.2f).

Oxidation number	Ion	Formula	Substance used
+7	persulphate ion,	$S_2O_8^{2-}$	potassium persulphate
+6	sulphate ion,	SO_4^{2-}	sodium sulphate
+5			
+4	sulphite ion,	SO_3^{2-}	sodium sulphite
+3			
+2	thiosulphate ion,	$S_2O_3^{2-}$	sodium thiosulphate
+1			
0	sulphur,	S_8	sulphur
−1			
−2	sulphide ion,	S^{2-}	sodium sulphide

Figure 19.2f

$E^{\ominus}$ values may help us to predict the stability of some of these states and how they may react with other substances. Figures 19.2g and 19.2h give these values at pH = 0 and pH = 14.

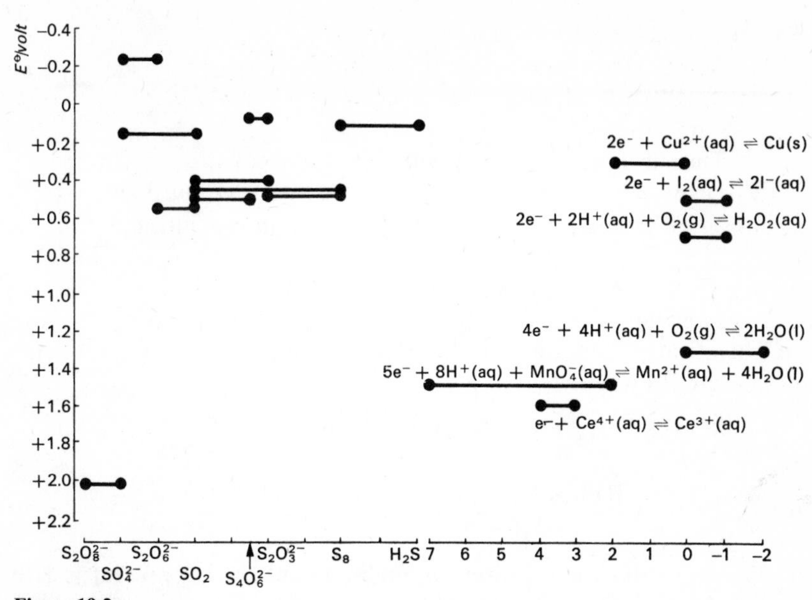

Figure 19.2g
Redox potentials at pH = 0.

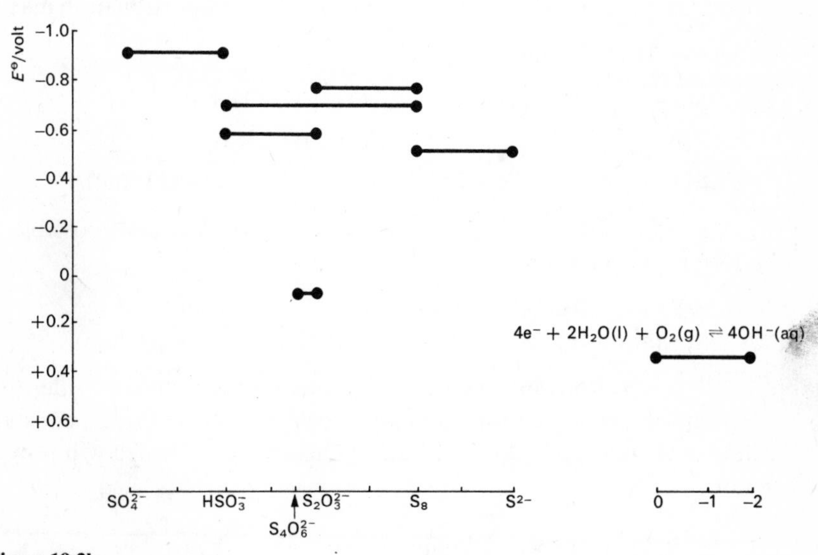

Figure 19.2h
Redox potentials at pH = 14.

Experiment 19.2a

Reactions of some sulphur compounds

If acid or alkali is added to substances which contain sulphur, two reactions may occur:

1 The substance may react with the acid or alkali.

2 Disproportionation may occur, that is, the substance oxidizes and reduces itself as nitrite ion does in acid solution to give nitrate ion and nitrogen oxide.

Sulphite

In a solution of sulphite ions, there is an equilibrium which can be thought of in a simplified way as:

$$SO_2(aq) + H_2O(l) \rightleftharpoons HSO_3^-(aq) + H^+(aq)$$

for which $K_c = \dfrac{[HSO_3^-]\,[H^+]}{[SO_2]} \approx 10^{-2}$

Therefore in alkaline solution the predominant species present is $HSO_3^-(aq)$, and in acid solution $SO_2(aq)$. If we warm the acid solution, unless something else happens, $SO_2(g)$ may be formed as the solubility of any gas in a liquid decreases as the temperature rises. Alternatively disproportionation may occur.

One possibility in acid solution is:

$$2SO_2(aq) + 2H^+(aq) + 4e^- \rightarrow S_2O_3^{2-}(aq) + H_2O(l)$$
$$[SO_2(aq) + 2H_2O(l) \rightarrow SO_4^{2-}(aq) + 4H^+(aq) + 2e^-] \times 2$$
$$4SO_2(aq) + 3H_2O(l) \rightarrow 2SO_4^{2-}(aq) + S_2O_3^{2-}(aq) + 6H^+(aq)$$

From the redox potential chart we can see that this reaction is possible, as are other disproportionations:

$$3SO_2(aq) + 2H_2O(l) \rightarrow 2SO_4^{2-}(aq) + S(s) + 4H^+(aq)$$

and $$7SO_2(aq) + 4H_2O(l) \rightarrow 3SO_4^{2-}(aq) + S_4O_6^{2-}(aq) + 8H^+(aq)$$

but as these equations involve hydrogen ions on the righthand side, the concentrations of which are raised to high powers in the equilibrium constant, the reactions are very sensitive to $[H^+(aq)]$, a high value of which will prevent the reaction.

1 How do you expect sulphite ions to react with acid?

a Test your prediction by warming some solid sodium sulphite with bench dilute hydrochloric acid.

Under alkaline conditions, the equilibrium:

$$SO_2(aq) + H_2O(l) \rightleftharpoons HSO_3^-(aq) + H^+(aq)$$

will obviously be shifted to the right so SO_2 will not be given off. The disproportionations in alkaline solution are:

$$4HSO_3^-(aq) + 2OH^-(aq) \rightarrow S_2O_3^{2-}(aq) + 2SO_4^{2-}(aq) + 3H_2O(l)$$
$$3HSO_3^-(aq) + OH^-(aq) \rightarrow S(s) + 2SO_4^{2-}(aq) + 2H_2O(l)$$

2 From the $E^\ominus$ values, are these reactions possible?
3 What effect will increasing the $[OH^-(aq)]$ have?
4 If sulphur is formed, it will be easily recognizable. How can you find out if sulphate has been formed?

b Add a few cm^3 of bench sodium hydroxide to some sodium sulphite in a test-tube and warm.
5 Has either of the reactions occurred? Why do you think this is so?

Sulphate
6 Is disproportionation possible? Why?

Thiosulphate
7 What disproportionations can occur in acid solution? (Refer to $E^\ominus$ values.)

c Warm some solid sodium thiosulphate with some bench dilute hydrochloric acid.
8 Which disproportionation has occurred?
9 Which disproportionations could occur in alkaline solution?

Sulphide
10 Is disproportionation possible? Why?

d Warm some solid sodium sulphide with bench sodium hydroxide solution.

e Warm some solid sodium sulphide with bench dilute hydrochloric acid.
11 Account for your observations.

Sulphur

12 Can sulphur disproportionate in acid solution?

13 Can sulphur disproportionate in alkaline solution?

f Add a few cm^3 bench sodium hydroxide solution to a few *small* pieces of roll sulphur. Boil for a few minutes, cool and add twice as much bench dilute hydrochloric acid. Account for your observations.

Thermal stability of various compounds of sulphur

If we heat one of these compounds, it may decompose. If it does, it will be because one or more of the bonds have broken. Bond energy terms should give us some information about this, and values for these for some bonds involving sulphur atoms are as follows:

Bond	Bond energy term/$kJ\ mol^{-1}$
S—S	251
S—H	344
S—O	435
O—O (peroxy)	146

Compared with C—C ($346\ kJ\ mol^{-1}$) and O—H ($463\ kJ\ mol^{-1}$) which we know are reasonably stable, we can see that molecules or ions which contain S—S or O—O bonds will probably decompose on heating.

14 Which ions or molecules in the following list contain S—S or O—O bonds and so may decompose on heating:

persulphate	sulphur
sulphate	sulphide
sulphite	thiosulphate?

g Heat about 0.1 g of each of these substances mentioned in question 14, and find out if your predictions are confirmed.

Experiment 19.2b

The oxidation of sulphite ions in solution

The oxidation of sulphite ions by iodine, cerium(IV) ions, and permanganate ions will be studied.

From $E^{\ominus}$ values it can be seen that it is possible for all these oxidizing agents to oxidize sulphite through dithionate to sulphate. In practice, the reaction may be kinetically controlled, and only dithionate may be formed, or a mixture of both dithionate and sulphate may be obtained.

The amount of oxidant consumed by 1 mole of sulphite ions will indicate whether sulphate only, sulphate and dithionate, or dithionate only has been formed.

1 Iodine

Pipette 25 cm^3 of 0.05M iodine solution into a 250 cm^3 conical flask and add approximately 5 cm^3 of bench dilute sulphuric acid. Titrate with 0.1M sodium sulphite solution slowly, continually swirling the flask. The end point is when the solution becomes colourless. Calculate how many moles of iodine reacted with 1 mole of sulphite ions and hence to what product the sulphite is oxidized.

2 Cerium(IV)

Pipette 25 cm^3 of 0.1M cerium(IV) sulphate solution into a conical flask. This solution contains sulphuric acid. Titrate with the sulphite solution. The end point is again when the solution becomes colourless. Calculate to what product the sulphite is oxidized in this case.

3 Permanganate

Pipette 25 cm^3 of 0.02M potassium permanganate solution into a conical flask and add about 10 cm^3 of bench dilute sulphuric acid. Titrate with the sulphite solution until the solution is colourless. Again calculate to what product the sulphite is oxidized.

Index

References to specific salts are indexed under the name of the appropriate cation; references to substituted organic compounds are indexed under the name of the parent compound.